CHEMISTRY IN FOCUS

SKILLS AND ASSESSMENT
WORKBOOK

YEAR 12

Debra Smith

Jaya Chowdhury

Samantha Dreon

Chemistry in Focus: Skills and Assessment Workbook Year 12
1st Edition
Debra Smith
Jaya Chowdhury
Samantha Dreon
ISBN 9780170449656

Publisher: Sam Bonwick
Editorial manager: Simon Tomlin
Editor: Marcia Bascombe
Proofreader: Catherine Greenwood
Original cover design by Chris Starr (MakeWork), Adapted by : Justin Lim
Text design: Ruth Comey (Flint Design)
Project designer: Justin Lim
Permissions researcher: Wendy Duncan
Production controller: Alice Kane
Typeset by: MPS Limited

Any URLs contained in this publication were checked for currency during the production process. Note, however, that the publisher cannot vouch for the ongoing currency of URLs.

Acknowledgements
Cover image: iStock.com/ivstiv
Inquiry questions on pages 5, 15, 18, 26, 52, 60, 72, 97, 104, 112, 120, 138, 165, 178 and 194 are from the Chemistry Stage 6 Syllabus © NSW Education Standards Authority for and on behalf of the Crown in right of the State of New South Wales, 2017.

For product information and technology assistance,
in Australia call **1300 790 853**;
in New Zealand call **0800 449 725**

For permission to use material from this text or product, please email **aust.permissions@cengage.com**

ISBN 978 0 17 044965 6

Cengage Learning Australia
Level 7, 80 Dorcas Street
South Melbourne, Victoria Australia 3205

Cengage Learning New Zealand
Unit 4B Rosedale Office Park
331 Rosedale Road, Albany, North Shore 0632, NZ

For learning solutions, visit **cengage.com.au**

Printed in China by 1010 Printing International Limited.
1 2 3 4 5 6 7 25 24 23 22 21

CONTENTS

ABOUT THIS BOOK v

MODULE FIVE » EQUILIBRIUM AND ACID REACTIONS — 1

Reviewing prior knowledge 1

1 Static and dynamic equilibrium — 5

Worksheet 1.1: Analysing the reversibility of chemical reactions. 5
Worksheet 1.2: Identifying static and dynamic equilibrium 8
Worksheet 1.3: Analysing examples of non-equilibrium systems in terms of entropy and enthalpy. 9
Worksheet 1.4: Analysing equilibrium graphs. 11

2 Factors that affect equilibrium — 15

Worksheet 2.1: Investigating dynamic equilibrium and factors that affect it 15

3 Calculating the equilibrium constant (K_{eq}) — 18

Worksheet 3.1: Interpreting equilibrium constants. . . 18
Worksheet 3.2: Applying ICE tables 20
Worksheet 3.3: Calculating the equilibrium constant 23

4 Solution equilibria — 26

Worksheet 4.1: Measuring and analysing solubility. 26
Worksheet 4.2: Applying solubility rules 30
Worksheet 4.3: Investigating solubility products 34
Worksheet 4.4: Problems solving solubility products. 37
Module five: Checking understanding. 42

MODULE SIX » ACID–BASE REACTIONS — 48

Reviewing prior knowledge 48

5 Properties of acids and bases — 52

Worksheet 5.1: Using indicators 52
Worksheet 5.2: Applying neutralisation reactions. 55
Worksheet 5.3: Defining acids and bases 58

6 Using Brønsted–Lowry theory — 60

Worksheet 6.1: Calculating pH and pOH 60
Worksheet 6.2: Comparing concentration and strength of acids. 63
Worksheet 6.3: Calculating pH and pOH of solutions 66
Worksheet 6.4: Using dissociation constants. 69

7 Quantitative analysis — 72

Worksheet 7.1: Analysing titrations 72
Worksheet 7.2: Using titration curves. 79
Worksheet 7.3: Exploring other titration techniques. 84
Worksheet 7.4: Applying buffers. 87
Module six: Checking understanding. 89

MODULE SEVEN » ORGANIC CHEMISTRY 94

Reviewing prior knowledge 94

8 Nomenclature 97

Worksheet 8.1: Investigating IUPAC nomenclature of hydrocarbons and haloalkanes................................. 97
Worksheet 8.2: Investigating IUPAC nomenclature of functional groups............ 100
Worksheet 8.3: Distinguishing isomers 102

9 Hydrocarbons 104

Worksheet 9.1: Investigating properties within a homologous series 104
Worksheet 9.2: Examining uses and disposal of organic substances 108

10 Products of reactions involving hydrocarbons 112

Worksheet 10.1: Investigating addition reactions.................................... 112
Worksheet 10.2: Investigating substitution reactions.................................... 115
Worksheet 10.3: Distinguishing between a saturated and an unsaturated hydrocarbon 117

11 Alcohols 120

Worksheet 11.1: Investigating alcohols 120
Worksheet 11.2: Investigating enthalpy of combustion of alcohols..................... 124
Worksheet 11.3: Predicting products of reactions of alcohols 127
Worksheet 11.4: Describing production of alcohols 130
Worksheet 11.5: Investigating oxidation of alcohols 132
Worksheet 11.6: Comparing fuels 135

12 Reactions of organic acids and bases 138

Worksheet 12.1: Investigating properties of functional groups........................ 138
Worksheet 12.2: Investigating esters................ 142
Worksheet 12.3: Investigating soaps and detergents.............................. 145

13 Polymers 148

Worksheet 13.1: Investigating addition polymers ... 148
Worksheet 13.2: Investigating condensation polymers...................................... 151
Module seven: Checking understanding 154

MODULE EIGHT » APPLYING CHEMICAL IDEAS 161

Reviewing prior knowledge 161

14 Analysis of inorganic substances 165

Worksheet 14.1: Investigating precipitation......... 165
Worksheet 14.2: Investigating ions in solution 167
Worksheet 14.3: Applying quantitative analysis techniques.................................... 170
Worksheet 14.4: Applying instrumental analysis techniques.................................... 174

15 Analysis of organic substances 178

Worksheet 15.1: Investigating organic compounds.................................... 178
Worksheet 15.2: Applying mass spectroscopy techniques.................................... 180
Worksheet 15.3: Applying NMR spectroscopy techniques.................................... 183
Worksheet 15.4: Applying IR and UV–visible spectroscopy techniques....................... 186
Worksheet 15.5: Combining spectroscopy techniques.................................... 189

16 Chemical synthesis and design 194

Worksheet 16.1: Examining synthesis reactions..... 194
Worksheet 16.2: Determining yield and purity...... 198
Module eight: Checking understanding 202

PRACTICE EXAMINATION 211
ANSWERS 226

9780170449656

ABOUT THIS BOOK

FEATURES

▶ Questions are provided to review prior knowledge from Year 11 at the start of each module and check your understanding of key concepts at the end of each module.

▶ Learning goals are stated at the top of each worksheet to set the intention and help you understand what's required.

▶ Chapters clearly follow the sequence of the syllabus and are organised by inquiry question.

▶ Page references to the content-rich student books provide an integrated learning experience.

▶ Brief content summaries are provided where applicable.

▶ Hint boxes provide guidance on how to answer questions effectively.

▶ A complete practice exam is provided.

▶ Fully worked solutions appear at the back of the book to allow you to work independently and check your progress.

ORGANISATION OF YOUR WORKBOOK

Each chapter begins with the relevant inquiry question and follows the sequence of the syllabus. Worksheets have been designed to complement the student book and provide additional opportunities to apply and revise your learning. Completion of these worksheets will provide you with a solid foundation to complete assessments and depth studies.

The workbook ends with a practice exam for you to complete independently. Use the solutions and marking criteria in the answers section to self-evaluate and plan for improvement.

EQUILIBRIUM AND ACID REACTIONS

Reviewing prior knowledge

1 For each statement, write true or false. Provide a reason for your response for statements that are false.

a When solid sodium hydroxide dissolves in water, the enthalpy increases as indicated by a temperature increase.

_____

b Respiration is an endothermic process.

_____

c Photosynthesis is an endothermic process.

_____

_____

d The precipitation of silver chloride results in an increase in entropy.

_____

_____

e The reaction $N_2(g) + 3H_2(g) \rightarrow 2NH_3(g)$ results in a decrease in entropy.

_____

_____

f Gibbs free energy measures the spontaneity of a reaction at 25°C.

_____

g A chemical reaction occurs when the particles have lots of energy.

_____

_____

h The activation energy is the minimum energy required for a reaction to occur.

_____

_____

i The activation energy for the forward reaction is always less than that for the reverse reaction in an exothermic reaction.

_____

_____

_____

j The activation energy for the forward reaction is always less than that for the reverse reaction in an endothermic reaction.

_____

_____

_____

k A catalyst speeds up the rate of a reaction by increasing the activation energy.

_____

_____

l A saturated solution is a solution in which no more solute can dissolve.

_____

m Temperature is a measure of the kinetic energy of molecules.

_____

2 The rate of reaction between calcium carbonate and nitric acid can be increased by changing several factors. The reaction is exothermic.

a List three factors that would increase the reaction rate.

_____

_____

_____

_____

b Write a balanced equation for the reaction of calcium carbonate with nitric acid.

_____

c Draw an energy profile diagram to show the pathways for the reaction in part b with and without a catalyst. Include activation energy, E_a and ΔH in your diagram.

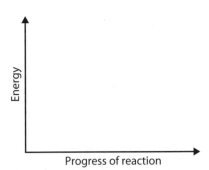

d The graph below shows the volume of carbon dioxide produced during the reaction between $1\,mol\,L^{-1}$ nitric acid and calcium carbonate.

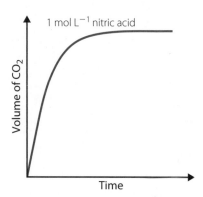

i Explain the shape of the curve.

_____

_____

_____

_____

_____

_____

_____

_____

_____

_____

_____

ii Draw a curve on the graph above to show the rate of reaction using $0.5\,mol\,L^{-1}$ nitric acid.

3 a Write a balanced equation to show the complete combustion of one mole of gaseous propane, C_3H_8, to produce carbon dioxide and steam. State whether ΔH is negative or positive for the reaction.

_____

b Explain whether there is an increase or a decrease in entropy in the above reaction.

_____

_____

_____

4 When solid ammonium chloride dissolves in water, the temperature of the water decreases.

a Write an equation to show the dissolving process.

_____

b Describe the changes that occur in bonding, enthalpy and entropy for the above process. Include a suitable labelled diagram in your response.

_____

_____

_____

_____

_____

_____

_____

_____

_____

5 The Gibbs free energy for the reaction shown is given for the temperatures stated.

$$CH_4(g) + H_2O(g) \rightarrow CO(g) + 3H_2(g)$$

At 500 K, $\Delta G^\theta = +98\,kJ\,mol^{-1}$

At 1200 K, $\Delta G^\theta = -53\,kJ\,mol^{-1}$

Explain at which temperature the reaction would be spontaneous.

_____

_____

6 Sketch a graph to show how the concentration of the reactant, A, and product, B, change for the reaction A $\rightarrow$ B, from the start to the end of reaction. Explain the shape of your graph.

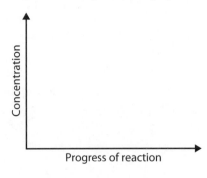

_____

_____

_____

1 Static and dynamic equilibrium

INQUIRY QUESTION: WHAT HAPPENS WHEN CHEMICAL REACTIONS DO NOT GO THROUGH TO COMPLETION?

WS 1.1 Analysing the reversibility of chemical reactions

STUDENT BOOK
Pages 28–34

LEARNING GOALS

Design an experiment to analyse whether a reaction is reversible

Determine whether a reaction is exothermic or endothermic

Write equations to represent a reaction at equilibrium

Identify species that change the position of equilibrium

Predict the colour of reactions at equilibrium based on colours of all species

1 A chemist wanted to investigate the reversibility of the cobalt(II) chloride hydrated and dehydrated system. Cobalt(II) chloride hexahydrate, $CoCl_2.6H_2O$, is a pink solid, while cobalt(II) chloride, $CoCl_2$, is a blue solid.

   a Outline an experiment to investigate the reversibility of this hydrated–dehydrated system.

   _____

   _____

   _____

   _____

   _____

   _____

   _____

   _____

   b Record the observations from the experiment in the table below.

Compound	Observation
Initial	
After heating	
After cooling	

   c Explain whether the observed reaction is reversible.

   _____

   _____

d State one check that could be performed on the pink solid obtained after cooling to prove it is the cobalt(II) chloride hexahydrate.

_____

_____

2 The iron(III) nitrate and potassium thiocyanate equilibrium system was investigated by a group of students who were investigating aqueous systems.

a Write a net ionic equation for this equilibrium system.

_____

b State the colour of the above equilibrium system.

_____

c State the colour of each of the species in the table below.

Species	Colour
Fe^{3+}	
SCN^-	
$FeSCN^{2+}$	

d A series of experiments was conducted to determine the effects of applied changes on the reversibility of the above equilibrium.

i An equilibrium system with an orange colour was placed in hot water and the colour faded, while in cold water it became a darker red colour. Is the equilibrium reaction as written above exothermic or endothermic? Explain.

_____

_____

_____

_____

_____

_____

_____

_____

_____

_____

_____

_____

_____

_____

ii The students added a colourless solution to the equilibrium mixture that caused the colour to fade and a yellow-brown precipitate to form. Suggest a solution that would cause the described change. Justify your choice with an equation.

_____

_____

_____

_____

_____

_____

_____

iii The students added a colourless solution to the equilibrium mixture that caused the colour to become more red. Suggest a solution that would cause the described change. Justify your choice.

_____

_____

_____

_____

_____

LEARNING GOALS

Write an equation for a system at equilibrium

Identify signs of enthalpy, entropy and Gibbs free energy at equilibrium

Distinguish between static and dynamic equilibrium

1 Nitrogen gas and hydrogen gas react to form ammonia gas and attain equilibrium in an exothermic reaction.

   a Write a balanced equation for the reaction at equilibrium.

   _____

   b The reaction written in part a is reversible. Use the particles given to show the species present before and after reaction in the boxes below.

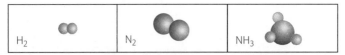

Before reaction	After reaction

   c For the above reaction, complete the table.

	>0, ~0 or <0
ΔH	
ΔS	
ΔG	

2 To distinguish between static and dynamic equilibrium, a student drew two diagrams as shown and labelled the flasks W and X. There was a liquid that partially evaporated at room temperature in both flasks but flask W was stoppered, while flask X was left open. The student's description is given below.

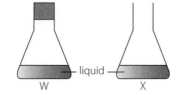

Static equilibrium is represented by flask W because there is nothing happening and the flask is sealed. There are no macroscopic changes.
The liquid just stays as a liquid because it cannot evaporate and escape the flask by turning into a gas, then come back into the flask again and condense into a liquid. Dynamic equilibrium is shown by flask X because it is open and the liquid can evaporate and leave the flask, and come back again into the flask and condense into a liquid. This process of the liquid turning into a gas leaving and re-entering the flask shows a dynamic equilibrium.

Rewrite the student's explanation, making appropriate amendments.

_____

_____

_____

_____

_____

_____

WS 1.3 Analysing examples of non-equilibrium systems in terms of entropy and enthalpy

STUDENT BOOK
Pages 25–38

LEARNING GOALS

Predict the sign of the enthalpy and entropy drives of a reaction

Write balanced symbolic equations for word equations

Compare the effects of enthalpy and entropy on reaction spontaneity

Recognise the role of activation energy in a chemical reaction

Apply collision theory to rates of reactions

1 Complete the table for the reactions given.

	Reaction	Balanced equation	ΔH (+ or –)	ΔS (+ or –)
a	Carbon monoxide gas reacts with oxygen gas to produce carbon dioxide gas. Heat is released.			
b	Liquid octane, C_8H_{18}, reacts with oxygen gas to produce carbon dioxide and steam. Heat is released.			
c	Solid glucose, $C_6H_{12}O_6$, reacts with oxygen to produce carbon dioxide gas and water vapour. Heat is released.			
d	Carbon dioxide and water react in the presence of chlorophyll and sunlight to produce solid glucose and oxygen gas. Heat is absorbed.			

2 a Write an equation for the complete combustion of liquid ethanol, CH_3CH_2OH, to produce gaseous products. The reaction releases heat.

_____

b Ethanol can be used as a fuel because its combustion after ignition is referred to as a spontaneous process. Explain the need for a spark or flame to initiate the combustion and why it is a spontaneous process once the reaction starts.

_____

_____

_____

_____

_____

_____

_____

3 When solid sodium nitrate dissolves in water, the temperature of the water decreases.

a Write an equation for the dissolution of sodium nitrate.

_____

b State the sign of enthalpy and entropy drives for the above dissolution.

_____

c Predict which drive is greater – enthalpy or entropy – given that the dissolution of sodium nitrate occurs as written.

_____

_____

_____

4 a Label the graph below using the following terms: activated complex, activation energy, reactants, products and ΔH.

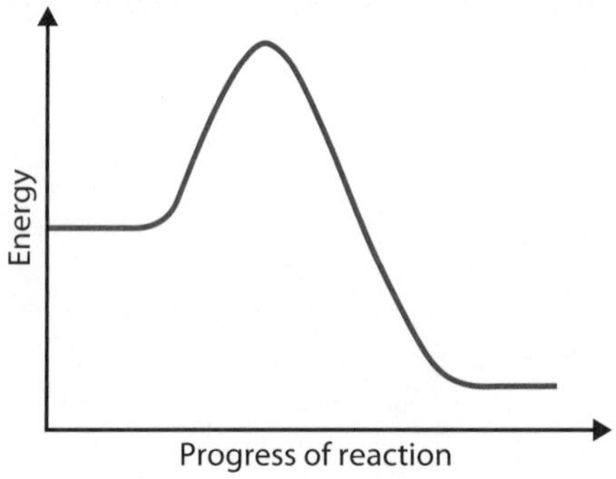

Energy (vertical axis) vs Progress of reaction (horizontal axis)

b i The reaction in part a can be represented as a general reaction of A + B → C. Justify whether the reaction is exothermic or endothermic.

_____

ii Describe the effect of increasing the temperature on the rate of the reaction in part bi in terms of collision theory.

_____

_____

_____

_____

c Predict the effect of activation energy on whether a reaction is reversible.

_____

_____

_____

_____

WS 1.4 Analysing equilibrium graphs

STUDENT BOOK
Pages 39–48

Describe a reaction from a graph of concentration versus time

Analyse graphs to predict changes made to equilibrium systems

Draw graphs to represent equilibrium systems and effects of changes

Distinguish between reaction rate versus time and concentration versus time graphs

1 The graph shows the changes in concentration of three species, labelled W, X and Y, in a reaction vessel over time.

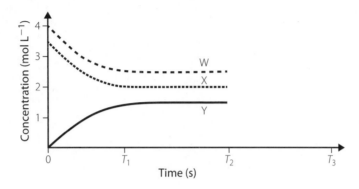

a Write an equation for the reaction and explain your response.

_____

_____

_____

_____

_____

_____

_____

_____

_____

_____

b In which direction does the equilibrium lie?

_____

c At time T_2, more W and X are added to the system and the system reaches equilibrium again a short time after T_2. Sketch the changes that occur on the graph above.

2 The graph below shows an important equilibrium reaction in the production of sulfuric acid.

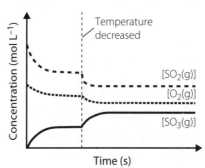

a Write the equilibrium reaction shown above. Explain whether your equation as written is endothermic or exothermic.

_____

_____

_____

_____

_____

b At time marked X in the graph below, the volume of the reaction vessel was decreased. Sketch on the graph below the changes that occur and explain your graph.

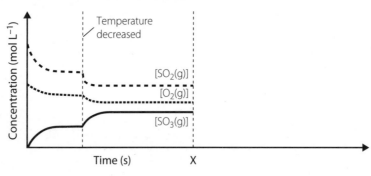

_____

_____

_____

_____

_____

_____

_____

_____

3 State the identity of X, which is added at time T_1 to a reaction. Give a reason for your response with reference to collision theory.

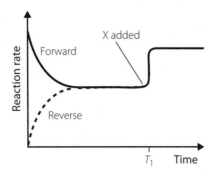

_____

_____

_____

_____

_____

_____

_____

_____

4 An example of a homogeneous reversible reaction is shown below.

$$2A(g) + B_2 \rightleftharpoons 2AB(g) \quad \Delta H = -200 \, \text{kJ mol}^{-1}$$

Read the description given and draw an appropriate graph with axes labelled to represent the changes. Explain your graph.

The system was at equilibrium between 0 and 2 minutes. The concentrations of A, B_2 and AB were 2, 4 and 5 mol L^{-1} respectively. At 2 minutes, the volume of the system was halved and equilibrium was re-established at 3 minutes. The temperature was increased at 5 minutes. A new equilibrium was established at 6.5 minutes and maintained until 8 minutes.

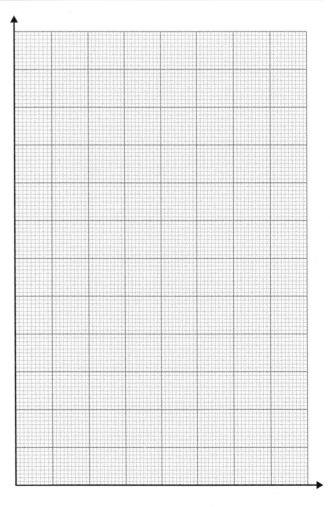

9780170449656

INQUIRY QUESTION: WHAT FACTORS AFFECT EQUILIBRIUM AND HOW?

WS 2.1 Investigating dynamic equilibrium and factors that affect it

STUDENT BOOK
Pages 34–49

LEARNING GOALS

Identify characteristics of dynamic equilibrium

Predict the colour of a system at equilibrium

Predict macroscopic change(s) when an equilibrium is affected by a factor

Apply Le Chatelier's principle and/or collision theory to an equilibrium system

1 Write true or false for the following statements that relate to dynamic equilibrium. If a statement is false, rewrite it so it is correct.

   a The macroscopic properties are not constant.

_____

   b The concentrations of reactants and products are the same.

_____

   c Dynamic equilibrium can only be attained in a closed system.

_____

   d When the reaction between nitrogen gas and hydrogen gas to produce ammonia reaches equilibrium, only ammonia gas is present in the reaction vessel.

_____

   e Equilibrium can be achieved in either direction for a reaction.

_____

2 In a sealed glass syringe, the following equilibrium was established at room temperature.

$$N_2O_4(g) \rightleftharpoons 2NO_2(g)$$

Note: N_2O_4 gas is colourless, while NO_2 gas is dark brown in colour.

   a Describe and explain the colour of the equilibrium mixture at room temperature.

_____

_____

_____

_____

b When the sealed syringe is placed in a beaker of hot water, the colour change observed is from brown to darker brown. Identify whether the equilibrium as written above is endothermic or exothermic. Explain your response.

_____

_____

_____

_____

_____

_____

_____

_____

c The equilibrium mixture was subjected to another change, but it was not a change in temperature, and this time the brown mixture became a lighter brown colour. Suggest the possible change that occurred, providing a reason for your response.

_____

_____

_____

_____

_____

_____

_____

3 Consider the reaction at equilibrium:

$$Cu(H_2O)_4^{2+}(aq) + 4Cl^-(aq) \rightleftharpoons CuCl_4^{2-}(aq) + 4H_2O(l)$$

The colours are: $Cu(H_2O)_4^{2+}$ is blue, $CuCl_4^{2-}$ is green and Cl^- is colourless.

a Describe and explain the colour of the equilibrium mixture.

_____

_____

_____

_____

b Explain what will happen if some dilute hydrochloric acid is added to the above system at equilibrium.

_____

_____

_____

_____

_____

_____

_____

_____

_____

_____

c Describe and explain how the blue-green equilibrium system above can be made a darker blue colour.

_____

_____

_____

> **HINT**
>
> Not by adding either product or reactant

_____

_____

_____

d Consider the heterogeneous equilibrium given below:

$$I_2(s) \rightleftharpoons I_2(g)$$

Note: Iodine solid is almost black in colour, while iodine gas is purple.

Complete the table.

Change	Observation	Effect of change	Explanation
Solid I_2 is added.			
Gaseous I_2 is added.			
Volume of the reaction container is decreased.			

Calculating the equilibrium constant (K_{eq})

STUDENT BOOK
Pages 54–9

 Interpreting equilibrium constants

LEARNING GOALS

Write balanced equilibrium equations

Write equilibrium constant expressions

Analyse how changing the equilibrium position of a system affects the equilibrium constant

Interpret graphs showing data from reactions

1 a Answer the following as true or false.

   i The concentration of a pure solid and a liquid can be omitted from the equilibrium constant as the concentration ratio of these is equal to 1. _____

   ii The equilibrium expression for the reaction $CH_4(g) + H_2O(g) \rightleftharpoons CO(g) + 3H_2(g)$ can be written as

$$K_{eq} = \frac{[CH_4][H_2O]}{[CO][H_2]}$$

_____

   iii A K_{eq} value higher than 1 means the equilibrium lies to the left. _____

   iv $Q > K$ means the reverse reaction must occur for equilibrium to be reached. _____

   v Increasing the concentration of the reactants will increase the equilibrium constant. _____

b Rewrite the statements identified as false so they are correct.

_____

_____

_____

_____

_____

_____

2 Haemoglobin is a protein within blood that contains iron. It is responsible for carrying oxygen to cells of the body. Each haemoglobin molecule can carry four oxygen molecules.

This reaction is shown as $Hb(aq) + 4O_2(aq) \rightleftharpoons Hb(O_2)_4(aq)$.

a Write an equilibrium expression for this reaction.

_____

_____

When a person is exposed to carbon monoxide gas, carbon monoxide replaces oxygen on the haemogloblin molecule, limiting the blood's ability to supply cells with oxygen.

b **i** Write the balanced chemical equation for this reaction.

_____

ii Write an equilibrium expression for this reaction.

_____

_____

To reverse the effects of carbon monoxide poisoning, a person must wear a mask supplied with pure oxygen for a period of time. Eventually, the carboxyhemoglobin will produce oxygenated haemoglobin and carbon monoxide will be released to be exhaled out of the lungs.

c **i** Write the balanced chemical equation for this reaction.

_____

ii Write an equilibrium expression for this reaction.

_____

_____

3 When would you expect to see water in an equilibrium expression?

_____

_____

4 Hydrogen chloride gas is produced exothermically from a reaction between hydrogen and chloride gases.

a Write a balanced equation and hence an equilibrium expression for this reaction.

_____

_____

b What would happen to the value of the equilibrium constant if heat is added to the system?

_____

_____

_____

_____

c Why does temperature affect the equilibrium constant?

_____

_____

_____

_____

d Write the equilibrium expression for the reverse reaction, in which hydrogen chloride is decomposed into hydrogen and chloride gases.

_____

_____

e What is the relationship between the K_{eq} of the forward reaction and the K_{eq} of the reverse reaction?

_____

_____

Apply ICE tables to the calculation of equilibrium constants of reactions

1 A sample of $0.500 \, \text{mol L}^{-1}$ of A is placed into a system with the reaction $2A(g) \rightleftharpoons 4B(g) + C(g)$.

At equilibrium, the concentration of A is $0.350 \, \text{mol L}^{-1}$.

Use an ICE table to determine K_{eq} for the following balanced general reaction.

	2A(g) ⇌	4B(g) +	C(g)
Initial concentration			
Change in concentration			
Equilibrium concentration			

a Enter the initial and final concentrations of A into the ICE table above.

b At the start of a reaction, the reactant concentration will be high, but the concentration of the product will be zero. Enter the value of 0.000 for the **I**nitial concentrations of 4B and C in the table above.

c The **C**hange in concentration can be determined by the molar ratio. As the reactant is converted into product, the concentration of reactant will decrease. We can call this $-x$. As the products are formed, their concentration increases, so the **C**hange in concentration can be expressed as $+x$. The coefficient determines the degree of change. What is the degree of change for the reactant A and products B and C?

_____

d Calculate the **C**hange in concentration for A and enter the number into the ICE table.

This number represents $-x$. Determine the **C**hange in concentration for B and C based on molar ratios and enter the numbers into the ICE table.

e Determine the **E**quilibrium concentration of B and C and enter the numbers into the ICE table.

f Write the equilibrium expression for this reaction.

_____

_____

g Calculate the K_{eq} of this reaction using the equilibrium concentration values from the ICE table.

_____

_____

2 Consider the following reaction between nitrogen monoxide and chlorine gas.

	2NO(g) +	Cl$_2$(g) ⇌	2NOCl(g)
Initial concentration	2.63	1.92	
Change in concentration			
Equilibrium concentration			1.96

a Complete the ICE table.

b Calculate the equilibrium constant.

_____

_____

3 When heated, solid ammonium chloride decomposes to form ammonia and hydrogen chloride gas.

	NH$_4$Cl(s) ⇌	NH$_3$(g) +	HCl(g)
Initial concentration			
Change in concentration		+1.42	
Equilibrium concentration			

a Complete the ICE table.

b Suggest why the first column of this ICE table should remain empty.

_____

c Calculate the equilibrium constant.

_____

_____

4 Construct an ICE table to determine the equilibrium constant when a sample of 9.21 mol L^{-1} hydrogen gas and 2.70 mol L^{-1} nitrogen gas are reacted to produce ammonia. The final concentration of the hydrogen gas was 1.35 mol L^{-1}.

5 Construct an ICE table to determine the equilibrium constant when 8.8 mol L^{-1} HCl(g) is reacted with a volume of oxygen gas to form chlorine gas and water vapour. At equilibrium, the concentration of oxygen gas has dropped by 1.25 mol L^{-1} to 6.3 mol L^{-1}.

6 Solid ammonium hydrogen sulfide (NH_4HS) decomposes to form ammonia gas and hydrogen sulfide gas (H_2S). A 1.80 mole sample of ammonium hydrogen sulfide was placed in a sealed 2.50 L container to reach equilibrium. At equilibrium, there were 0.418 moles of ammonia gas.

Construct an ICE table to calculate the equilibrium constant for this reaction.

7 Nitrogen oxides are air pollutants produced by the reaction of nitrogen gas and oxygen gas at high temperatures. A 0.089 mole quantity of oxygen gas is mixed with 0.036 mol L^{-1} nitrogen gas in a 2 L vessel. At equilibrium, 0.0126 mol L^{-1} of nitrogen dioxide is produced.

Construct an ICE table to calculate the equilibrium constant for this reaction.

8 When 0.40 moles of PCl_5 is heated in a 10.0 L container, an equilibrium is established in which 0.25 moles of Cl_2 is present.

Construct an ICE table to calculate the equilibrium constant for this reaction.

Calculate the equilibrium constant of a reaction

Determine the balanced equation of a reaction

Interpret graphs

Calculate the concentration of different species using the equilibrium constant

1 Calculate the magnitude of the equilibrium constant for each of the following examples at equilibrium.

a $COBr_2(g) \rightleftharpoons CO(g) + Br_2(g)$

given that, at equilibrium, $[COBr_2] = 0.15\,mol\,L^{-1}$, $[CO] = 0.17\,mol\,L^{-1}$, $[Br_2] = 0.17\,mol\,L^{-1}$.

_____

_____

_____

b $2SO_3(g) \rightleftharpoons 2SO_2(g) + O_2(g)$

given that, at equilibrium $[SO_3] = 0.37\,mol\,L^{-1}$, $[SO_2] = 0.25\,mol\,L^{-1}$, $[O_2] = 0.86\,mol\,L^{-1}$.

_____

_____

_____

c $SO_3(g) + H_2O(g) \rightleftharpoons H_2SO_4(l)$

given that, at equilibrium, $[SO_3] = 0.400\,mol\,L^{-1}$, $[H_2O] = 0.480\,mol\,L^{-1}$, $[H_2SO_4] = 0.600\,mol\,L^{-1}$.

_____

_____

_____

d $PCl_5(g) + H_2O(g) \rightleftharpoons 2HCl(g) + POCl_3(g)$

given that, at equilibrium in a 2 L vessel, $PCl_5 = 0.075\,mol$, $H_2O = 0.050\,mol$, $HCl = 0.750\,mol$, $POCl_3 = 0.500\,mol$.

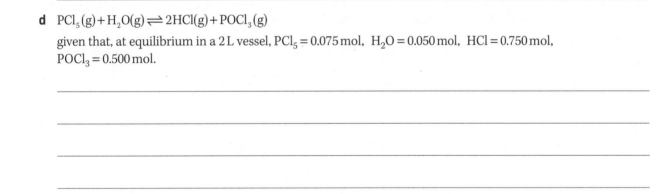

_____

_____

_____

_____

2 The following reaction took place within a 2.3 L vessel.

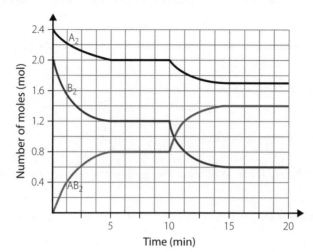

a Write a balanced equation for this reaction.

_____

b How long does it take the reaction to reach equilibrium for the first time?

_____

c Calculate the equilibrium constant at this point.

_____

_____

_____

_____

_____

_____

d At 10 minutes, a change is made to the system before it is allowed to return to equilibrium at 15 minutes. Calculate the equilibrium constant at this new equilibrium point.

_____

_____

_____

_____

_____

_____

e Comparing the K_{eq} calculations from parts c and d, explain the change that occurred in the system at 10 minutes.

_____

_____

f The K_{eq} value calculated in part c is said to be low. What does a low K_{eq} value indicate about the yield of a product in this reaction?

_____

_____

3 The concentrations of gases for the reaction $PCl_5(g) \rightleftharpoons PCl_3(g) + Cl_2(g)$ at 250°C were measured when the reaction was at equilibrium. The concentrations were:
$[PCl_5] = 4.2 \times 10^{-5}\,mol\,L^{-1}$, $[PCl_3] = 1.3 \times 10^{-2}\,mol\,L^{-1}$, $[Cl_2] = 3.9 \times 10^{-3}\,mol\,L^{-1}$.

a Calculate the equilibrium constant.

_____

_____

_____

b If the concentrations of PCl_3 and Cl_2 were both measured to be $1.0 \times 10^{-2}\,mol\,L^{-1}$, what would be the concentration of PCl_5?

_____

_____

_____

4 Ammonia is synthesised from nitrogen and hydrogen in the following reaction:

$$N_2(g) + 3H_2(g) \rightleftharpoons 2NH_3(g)$$

At 400°C, the equilibrium constant for this reaction is 0.080. If the concentration of N_2 is $0.600\,mol\,L^{-1}$ and that of H_2 is $0.420\,mol\,L^{-1}$, what will be the concentration of the ammonia produced?

_____

_____

_____

_____

5 Hydrofluoric acid dissociates into hydrogen ions and fluoride ions according to the following reaction:

$$HF(aq) \rightleftharpoons H^+(aq) + F^-(aq)$$

a Calculate the equilibrium constant at 25°C if $[H^+] = 0.0075\,mol\,L^{-1}$, $[F^-] = 0.0075\,mol\,L^{-1}$ and $[HF] = 0.085\,mol\,L^{-1}$.

_____

_____

_____

b Some hydrofluoric acid is added to water at 25°C and the hydrogen ion concentration is measured to be $0.010\,mol\,L^{-1}$. What is the equilibrium concentration of HF?

_____

_____

_____

_____

4 Solution equilibria

WS 4.1 Measuring and analysing solubility

STUDENT BOOK
Pages 85–7

LEARNING GOALS

Write balanced chemical equations

Describe changes in bonding, energy and entropy when a salt dissolves

Calculate enthalpy, entropy and Gibbs free energy

Analyse solubility curves

Use data to draw a solubility curve

Calculate solubility quantities

1 a Write a balanced equation for the dissolution of sodium chloride in water.

_____

b Describe the changes in bonding that occur when sodium chloride dissolves in water. Include a labelled diagram to support your answer.

_____

_____

_____

_____

_____

_____

_____

c Explain the changes in both energy and entropy that occur when sodium chloride dissolves.

_____

_____

_____

_____

_____

_____

_____

d Use the data given in the table below to calculate the following quantities for the dissolution of sodium chloride in water at 25°C.

Substance	ΔH^θ (kJ mol^{-1})	S^θ (J K^{-1} mol^{-1})
Na$^+$(aq)	−240	59.0
Cl$^-$(aq)	−167	56.5
NaCl(s)	−411	72.1

i Enthalpy. Comment on this value with reference to your answer to part c.

_____

_____

_____

_____

_____

_____

ii Entropy. Comment on this value with reference to your answer to part c.

_____

_____

iii ΔG^θ. Comment on the spontaneity of the reaction.

_____

_____

_____

_____

2 The solubility curves for various salts are shown below.

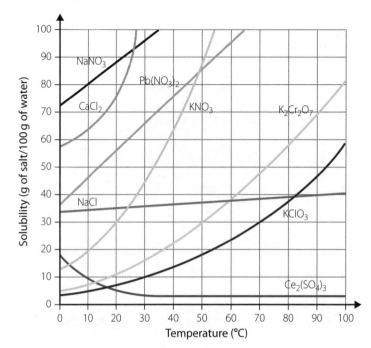

Use the solubility curves above to answer the following questions.

a **i** What mass of $KClO_3$ will dissolve in 50 g of water at 70°C?

_____

ii What mass of $CaCl_2$ will dissolve in 150 g of water at 10°C?

_____

b **i** How much KNO_3 will dissolve in 100 g of water at 10°C?

_____

ii How much more KNO_3 will dissolve at 50°C?

_____

_____

c A 40 g sample of $K_2Cr_2O_7$ is added to 100 mL of water and the temperature of the water is slowly increased.
   i At what temperature would the solution no longer be saturated?

_____

ii Above this temperature would the solution be saturated, unsaturated or supersaturated? Explain your answer.

_____

_____

d Use the information in the solubility curves above to state a generalised relationship between temperature and solubility, noting any exceptions to this statement.

_____

_____

_____

3 The table below gives the solubility of ammonium chloride in 100 g water at different temperatures.

Temperature (°C)	10	20	30	40	50	60	70	80	90
Solubility (g per 100 g water)	33	37	41	46	50	55	60	66	71

a Draw a graph of solubility versus temperature on the grid below.

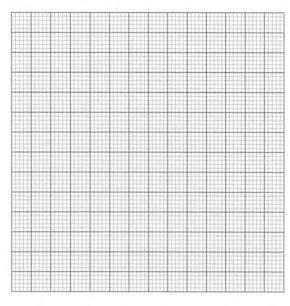

b Write a balanced equation to show the dissolution of NH_4Cl in water.

_____

c **i** How much NH_4Cl will dissolve in 100 g water at 35°C?

_____

ii Calculate the concentration of this solution in $mol\,L^{-1}$.

_____

_____

_____

_____

_____

_____

d What is the minimum mass of water at 65°C needed to dissolve 36 g of NH_4Cl?

_____

_____

1 A student was investigating mass and concentration relationships in a reaction between sodium chloride and silver nitrate.

In the experiment, different volumes of a known concentration of sodium chloride solution were added to separate 50 mL samples of a silver nitrate solution. After each reaction, the silver chloride precipitate was filtered off, dried and weighed.

The results of the experiment are below.

Volume of NaCl used (mL)	Volume of AgNO₃ used (mL)	Mass of precipitate (g)
10.0	50.0	0.145
20.0	50.0	0.290
30.0	50.0	0.440
40.0	50.0	0.580
50.0	50.0	0.720
60.0	50.0	0.720
70.0	50.0	0.715
75.0	50.0	0.725

a Write the net ionic equation for the reaction between $AgNO_3$ and NaCl.

_____

b Construct a graph to show the relationship between the mass of silver chloride formed (vertical axis) and volume of sodium chloride used (horizontal axis).

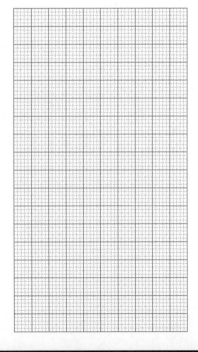

c Describe the graph.

_____

_____

d Explain why it changes shape.

_____

_____

e Calculate the number of moles of silver nitrate in 50 mL of the original solution.

_____

_____

f Calculate the molarity of the original silver nitrate solution.

_____

_____

2 Determine whether a precipitate is formed when the following solutions are mixed. If a precipitate does form, write the net ionic equation for the reaction.

a Sodium hydroxide and copper(II) nitrate

b Aluminium chloride and potassium carbonate

c Lead(II) nitrate and magnesium acetate

_____

_____

_____

3 A student was given four solutions and told each contained one of the following anions:

$$CH_3COO^-, O^{2-}, NO_3^-, SO_4^{2-}, Cl^-, CO_3^{2-}$$

The results of a series of tests that were conducted are given below.

Cation added	A	B	C	D		
Ba^{2+}	NP	NP	ppt	NP		
Ag^+	ppt	NP	ppt	ppt		
Ca^{2+}	NP	NP	ppt	NP		
Al^{3+}	NP	NP	NP	ppt		

NP – no precipitate; ppt – precipitate formed

Table 1 Solubility of common ionic substances

Soluble anions	Exceptions	Insoluble anions	Exceptions
NO_3^-	None	OH^-	Group 1, NH_4^+, Ba^{2+}, Sr^{2+} soluble; Ca^{2+} slightly soluble
CH_2COO^-	Ag^+ slightly soluble	O^{2-}	Group 1, NH_4^+, Ba^{2+}, Sr^{2+}, Ca^{2+} soluble
Cl^-	Ag^+ insoluble Pb^{2+} slightly soluble	S^{2-}	Groups 1 and 2, NH_4^+ soluble
Br^-	Ag^+ insoluble Pb^{2+} slightly soluble	CO_3^{2-}	Group 1, NH_4^+ soluble
I^-	Ag^+, Pb^{2+} insoluble	SO_3^{2-}	Group 1, NH_4^+ soluble
SO_4^{2-}	Ba^{2+}, Pb^{2+}, Sr^{2+} insoluble Ag^+, Ca^{2+} slightly soluble	PO_4^{3-}	Group 1, NH_4^+ soluble

a Use the solubility data in Table 1 and the given results to identify possible anions for A–D. Justify your choice.

_____

_____

_____

_____

_____

_____

_____

b Complete the results table on the previous page for the two anions that were not identified in your answer to part a.

4 Students were given a sample that they were told contained a mixture of lead nitrate, calcium nitrate and magnesium nitrate. They were asked to use the solubility rules to separate the cations. Draw a flow chart to show how the cations could be separated. Write a balanced net ionic equation for each reaction.

5 A small town was concerned about the effluent from a nearby factory, which was flowing into the local river. Water quality scientists tested the water and found it to contain lead ions.

To determine the concentration of lead in the water, they added 10 mL of 0.1 mol L^{-1} KI to a 100 mL sample of the water. The mixture was then filtered, and the precipitate was thoroughly dried and then weighed.

a Write a net ionic equation for the reaction.

_____

b The mass of the dried precipitate was 0.0154 g. Calculate the number of moles of lead ions in the sample.

_____

_____

c The limit for the maximum safe concentration of lead is 0.01 mg L^{-1}, as recommended by the Environmental Protection Authority.

Determine whether the quantity of lead in the river exceeds the safe limit.

_____

_____

_____

_____

d What assumption(s) did the scientists make and how would these affect the result if they were incorrect?

_____

_____

_____

_____

_____

_____

_____

_____

_____

Write balanced chemical equations

Derive equilibrium expressions

Determine an aim for an investigation

Analyse an experimental procedure

Apply Le Chatelier's principle

Identify relevant data

Perform stoichiometric calculations

Calculate K_{sp}

Suggest reasons for differences in theoretical and experimental values

1 After conducting background research for their depth study, a student decided they would conduct an investigation to determine the K_{sp} of the slightly soluble salt, calcium hydroxide. The student had already learnt that hydroxides were bases and these could be neutralised using an acid. They had also learnt that the acid–base indicator bromothymol blue could be used to show that neutralisation had been reached when the indicator changed from blue to green, and that when the indicator was yellow, too much acid had been added.

Using this information, the student decided to make a saturated solution of calcium hydroxide then use HCl to neutralise the OH^- ions present in solution. This would then be used to calculate the amount of dissolved OH^- present. This data would enable the student to calculate the K_{sp}.

a Write an aim for the investigation.

_____

b To begin, the student prepared a saturated solution of calcium hydroxide by adding approximately 1 g of calcium hydroxide to 100 mL distilled water. They allowed the solution to stand overnight. The next day the student filtered the solution, discarded the residue and kept the filtrate for analysis of hydroxide ions.

  i Suggest why they did not need to know the exact mass of calcium hydroxide added.

_____

_____

  ii The molar solubility of $Ca(OH)_2$ is $0.0108 \, mol \, L^{-1}$. Did the student add enough to ensure the solution was saturated?

_____

_____

_____

_____

  iii Write the equilibrium equation for the saturated solution.

_____

  iv With reference to your answer for part iii, explain why the solid had to be removed from the solution before the reaction with HCl could be conducted.

_____

_____

_____

_____

_____

_____

_____

_____

_____

_____

_____

_____

_____

c   In the next phase of the investigation, the student used a measuring cylinder to measure 10 mL of the filtrate into each of four separate conical flasks. To each of these they added two drops of the bromothymol blue acid–base indicator, which would change colour when neutralisation occurred. Then, using a burette containing $0.100 \, mol \, L^{-1}$ HCl, the student carefully added HCl to the first flask until the indicator changed colour. They recorded the volume of HCl added. They then repeated this process three more times. The data obtained is shown in the table below.

Flask	Volume HCl added (mL)
1	3.10
2	2.30
3	2.20
4	2.40

i   Calculate the average volume, explaining how you arrived at your answer.

_____

_____

_____

_____

_____

ii   Calculate the number of moles of HCl needed for neutralisation.

_____

_____

_____

iii   Calculate the number of moles and concentration of OH$^-$ in the solution.

_____

_____

_____

_____

d Using the value for OH⁻ from part c iii, calculate the concentration of Ca^{2+} ions in the solution.

_____

_____

_____

e Calculate the K_{sp} for $Ca(OH)_2$.

_____

_____

f Compare the experimental value obtained in part e with the theoretical value and suggest reasons for any difference.

_____

_____

_____

_____

Problems solving solubility products

LEARNING GOALS

Analyse accuracy of information

Predict solubility and calculate concentrations using K_{sp}

Identify trends in data and information

Calculate K_{sp} and predict precipitation using solubility and concentration

Predict the effects of changes to a system in equilibrium

Analyse the effect of a common ion on equilibrium concentrations of ions and precipitation

Justify identified assumptions

1 After reading the statement: 'Only insoluble and sparingly soluble salts reach dynamic equilibrium', a student asked their teacher the following question:

'Wouldn't a saturated solution of any ionic solid in water that contains excess solid be in dynamic equilibrium?'

Explain what you think the teacher's response should be.

_____

_____

_____

_____

_____

_____

_____

_____

_____

_____

_____

_____

_____

_____

_____

_____

_____

2 a Arrange the following group 2 hydroxides in order from most soluble to least soluble:
magnesium hydroxide, barium hydroxide and calcium hydroxide.

_____

b Identify any trend that exists.

_____

c Consider the group 2 carbonates and identify any trend that exists.

_____

_____

_____

d Explain whether a generalisation can be made about the solubility of different group 2 salts.

_____

_____

_____

_____

_____

_____

3 a The molar solubility of AgI is 2.11×10^{-7} g/100 g of water at 25°C. Calculate the K_{sp} for AgI.

_____

_____

_____

_____

_____

_____

_____

_____

b A 100 mL sample of a 1.47×10^{-9} mol L^{-1} solution of AgNO$_3$ is added to 100 mL of a 2.02×10^{-6} mol L^{-1} solution of MgI$_2$. Justify whether or not a precipitate forms.

_____

_____

_____

_____

_____

4 a The K_{sp} for calcium carbonate is 3.36×10^{-9}. Calculate the concentrations of calcium ions and carbonate ions in a saturated aqueous solution of calcium carbonate.

_____

_____

_____

_____

b Many aquatic organisms build shells made of $CaCO_3(s)$. At the surface of the ocean, the concentrations of $Ca^{2+}(aq)$ and $CO_3^{2-}(aq)$ are high enough for the calcium carbonate in the shells to be insoluble. However, no seashells are found deeper in the ocean where pressure is higher and temperature is lower. The K_{sp} of $CaCO_3$ is higher at greater depths.

Suggest reasons for this in terms of:

  i temperature (the dissolving of $CaCO_3$ is exothermic)

  ii K_{sp}

5 A solution that contains $0.125\ mol\ L^{-1}\ Pb^{2+}$ ions and $0.000\,750\ mol\ L^{-1}\ Ag^+$ ions has HCl slowly added to it. The K_{sp} of silver chloride is 1.77×10^{-10} and the K_{sp} of lead chloride is 1.70×10^{-5}.

  a Determine which precipitate forms first – $PbCl_2$ or AgCl. Show all calculations.

b Suggest a method that could be used to separate the lead and silver ions in 100 mL of the solution above.

_____

_____

_____

_____

_____

_____

_____

_____

_____

_____

_____

_____

_____

_____

_____

_____

6 a The K_{sp} of barium phosphate is 1.30×10^{-29} at 25°C. Calculate the concentration of barium ions and phosphate ions in a saturated aqueous solution of barium phosphate.

_____

_____

_____

_____

_____

_____

_____

_____

_____

b Given a saturated solution of barium phosphate, describe how each of the following changes could be brought about and explain why the change occurs.

  i Decrease in the concentration of barium ions

_____

_____

_____

_____

ii Increase in the concentration of phosphate ions

_____

_____

_____

_____

_____

_____

7 a How many grams of silver sulfate can be dissolved in 450 mL of 0.200 mol L^{-1} sodium sulfate solution? (K_{sp} of silver sulfate is 1.20×10^{-5}.)

_____

_____

_____

_____

_____

_____

_____

_____

_____

_____

_____

_____

_____

b Identify any assumptions made in the calculations in part a and justify their validity.

_____

_____

_____

_____

_____

_____

_____

Module five: Checking understanding

1 A system is said to have reached equilibrium when

  A all the reactants have been used up.

  B the rate of the reverse reaction is greater than the rate of the forward reaction.

  C the rate of the forward reaction is equal to the rate of the reverse reaction.

  D the rate of the reaction reaches zero.

2 What is the expression for K_{eq} for the following reaction?

$$2H_2O(g) \rightleftharpoons 2H_2(g) + O_2(g)$$

  A $K_{eq} = \dfrac{[2H_2O]}{[2H_2][O_2]}$

  B $K_{eq} = \dfrac{[H_2]^2[O_2]}{[H_2O]^2}$

  C $K_{eq} = \dfrac{[2H_2][O_2]}{[2H_2O]}$

  D $K_{eq} = \dfrac{[H_2O]^2}{[H_2]^2[O_2]}$

3 A mixture of SO_2, O_2, and SO_3 was placed in a 1.00 L vessel and allowed to reach equilibrium according to the equation

$$2SO_2(g) + O_2(g) \rightleftharpoons 2SO_3(g)$$

At equilibrium, the amount of each species was found to be 0.06 mol of SO_2, 0.06 mol of O_2, and 0.03 mol of SO_3. The magnitude of the equilibrium constant, K_{eq}, for this reaction is closest to which of the following?

  A 0.12

  B 0.24

  C 4.2

  D 8.3

4 Hydrogen iodide decomposes according to the following equation.

$$2HI(g) \rightleftharpoons H_2(g) + I_2(g)$$

The equilibrium constant for this reaction at 25°C is 0.0208.

Equimolar amounts of HI(g), $H_2(g)$ and $I_2(g)$ are mixed in a 1 L vessel at this temperature. Which one of the following statements is correct?

  A The concentration of H_2 would decrease.

  B The concentration of HI would decrease.

  C The value of K would increase to 1.

  D The pressure of the system would increase.

5 The activation energy of a reversible reaction decreases when

  A the temperature is decreased.

  B the surface area of a reactant is increased.

  C the concentration of the reactants is increased.

  D a catalyst is added.

6 Which of the following statements is correct?

The value of the equilibrium constant for a reaction changes when

  A the reaction begins.

  B the temperature is increased.

  C the concentrations of the reactants are increased.

  D a catalyst is added.

7 When solutions of KSCN and $FeCl_3$ are mixed, they react to produce $FeSCN^{2+}(aq)$, which gives the solution a red colour. The equation for the reaction is given below.

$$Fe^{3+}(aq) + SCN^-(aq) \rightleftharpoons FeSCN^{2+}(aq)$$

Which of the following would result in the intensity of the red colour being decreased?

A addition of $FeSCN^{2+}(aq)$

B addition of $SCN^-(aq)$

C addition of Ag^+ ions, which precipitate SCN^-

D addition of $Fe(NO_3)_3$

8 Equimolar solutions of the following were mixed. In which of the following would a precipitate be least likely to form?

A $AgNO_3$ and KI

B $CaCl_2$ and Na_2CO_3

C $MgCl_2$ and KNO_3

D $Ba(OH)_2$ and $FeSO_4$

9 Use the K_{sp} values to answer the following question.

For $BaCO_3$, 8×10^{-9}; for CaF_2, 3×10^{-1}; for $PbCl_2$, 2×10^{-4}; for AgI, 1×10^{-16}.

Which of the following statements is correct?

A $PbCl_2$ is the least soluble.

B $BaCO_3$ is more soluble than $PbCl_2$.

C AgI is the most soluble.

D CaF_2 is more soluble than $BaCO_3$.

10 A saturated solution of lead(II) hydroxide $(Pb(OH)_2)$ is prepared at 25°C. K_{sp} of lead(II) hydroxide at this temperature is 1.43×10^{-15}. What would be the concentrations of Pb^{2+} ions and OH^- ions in the solution?

	$[Pb^{2+}]\,[mol\,L^{-1}]$	$[OH^-]\,[mol\,L^{-1}]$
A	1.89×10^{-8}	3.78×10^{-8}
B	7.1×10^{-6}	7.1×10^{-6}
C	7.1×10^{-6}	1.42×10^{-5}
D	1.12×10^{-5}	2.24×10^{-5}

11 The system described by the equation below is in equilibrium. Explain, with reasons, three ways in which the equilibrium concentration of nitrogen monoxide, NO(g), could be increased.

$$4NH_3(g) + 5O_2(g) \rightleftharpoons 4NO(g) + 6H_2O(l) \qquad \Delta H = -908\,kJ\,mol^{-1}$$

_____

_____

_____

_____

_____

_____

_____

_____

_____

12 The vapour pressure of ethanol is 3.14 kPa at 10°C and 5.85 kPa at 20°C. If a closed system containing ethanol in equilibrium at 10°C was heated to 20°C, describe changes that would occur in the system.

_____

_____

_____

13 The concentrations of the three substances in the following reaction are shown in the graph below.

$$PCl_3(g) + Cl_2(g) \rightleftharpoons PCl_5(g) \qquad\qquad \Delta H = -93 \, kJ \, mol^{-1}$$

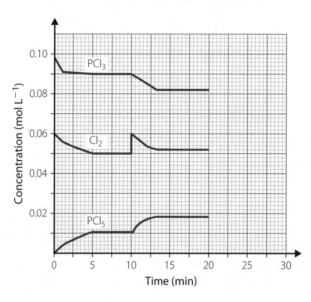

a Explain at what time the system first reached equilibrium.

_____

_____

b Suggest what happened to the system from 10 minutes to 20 minutes.

_____

_____

_____

_____

_____

_____

c On the graph, sketch an extension to each of the PCl_3 and PCl_5 lines to show what would happen if we _increased_ the volume of the whole system at 20 minutes until equilibrium is again established. Explain the reasoning behind your sketch.

_____

_____

_____

_____

14 Carbon dioxide, CO_2, the waste product of metabolism in body cells, enters the venous bloodstream from the cells and establishes the equilibrium system:

$$CO_2(aq) + H_2O(l) \rightleftharpoons H_2CO_3(aq) \rightleftharpoons HCO_3^-(aq) + H^+(aq) \quad \text{Reaction 1}$$

In the lungs, O_2 enters the arterial bloodstream and combines with haemoglobin, HbH^+, as follows:

$$HbH^+(aq) + O_2(g) \rightleftharpoons HbO_2(aq) + H^+(aq) \quad \text{Reaction 2}$$

The partial pressures of O_2 (PPO_2) and CO_2 ($PPCO_2$) at various locations in the body are shown in the table.

	Lungs	Body cells	Arterial blood	Venous blood
PPO_2 (atm)	0.137	0.046	0.125	0.059
$PPCO_2$ (atm)	0.053	0.066	0.053	0.059

Use Le Chatelier's principle to explain how the production of the waste product CO_2 by tissues and the concentration of O_2 in the lungs causes the two equilibria to shift in such a way as to assist the transport of O_2 from the lungs to the body cells.

15 Industrial ethanol can be produced by the acid-catalysed reaction between ethene and steam according to the equilibrium reaction

$$CH_2{=}CH_2(g) + H_2O(g) \rightleftharpoons CH_3CH_2OH(g)$$

When the reaction has reached equilibrium, in a 5.0 L vessel at 150°C, there are 0.50 mol C_2H_4, 1.0 mol H_2O and 0.030 mol CH_3CH_2OH present.

a Write the equilibrium expression for this reaction.

b Determine the value of the equilibrium constant at this temperature.

_____

_____

_____

_____

16 Phosgene, $COCl_2$, which is used in the synthesis of polycarbonate and polyurethane resins, is prepared by passing a mixture of carbon monoxide and chlorine over activated carbon.

During its production, the following equilibrium is established:

$$CO(g) + Cl_2(g) \rightleftharpoons COCl_2(g)$$

In an industrial laboratory investigation, equilibrium was established at a particular temperature in a 1.0 L vessel. On analysis, the equilibrium mixture was shown to contain 1.5 mol phosgene, 0.15 mol carbon monoxide and 1.0 mol chlorine. In a different 1.0 L vessel, at the same temperature, another equilibrium mixture contained 0.20 mol chlorine and 0.10 mol carbon monoxide. What mass of phosgene was present at equilibrium in the second flask?

_____

_____

_____

_____

_____

_____

_____

_____

_____

17 Barium hydroxide is shaken with water at 25°C until no more dissolves. The concentration of barium ions in the saturated solution is $0.108 \, mol \, L^{-1}$.

a Write the equation for the equilibrium reaction in the saturated solution.

_____

b Calculate the equilibrium constant for this reaction.

_____

_____

_____

_____

c Barium hydroxide is dissolved in a sodium hydroxide solution at 25°C and analysis shows that the concentration of the hydroxide ions in the saturated solution is $0.225 \, mol \, L^{-1}$. What is the concentration of barium ions in the solution?

_____

_____

_____

_____

_____

d A 50 mL sample of $0.300 \, mol \, L^{-1}$ sodium hydroxide is mixed with 50 mL of $0.210 \, mol \, L^{-1}$ barium chloride solution. Will a precipitate form?

_____

_____

_____

_____

_____

MODULE SIX »
ACID–BASE REACTIONS

Reviewing prior knowledge

1 What is the general reaction for the neutralisation of an acid and a base?

_____

2 Write the balanced chemical equations for the neutralisation reactions between the listed acids and bases.

   a Hydrochloric acid and sodium hydroxide

_____

   b Carbonic acid and lithium hydroxide

_____

   c Phosphoric acid and calcium hydroxide

_____

   d Hydrobromic acid and barium hydroxide

_____

   e Nitric acid and zinc hydroxide

_____

   f Acetic acid and aluminium hydroxide

_____

3 Identify the properties of an acid and base by completing the following table.

Property	Acid	Base
Taste		
Texture (feel)		
Change in litmus		
Possible ions involved		

4 Complete the following table with the correct formula that would be applied for each scenario, and then complete the calculation.

> **HINT**
>
> Don't forget to include units in your final calculation.

Subject	Formula to use	Calculation
Determine the number of moles in a 2.4 g sample of barium hydroxide.		
Calculate the mass of sodium hydroxide required to make a solution containing 0.135 moles.		
Determine the number of moles of hydrochloric acid in 22 mL of a 0.240 mol L^{-1} solution.		

Subject	Formula to use	Calculation
Calculate the concentration of a sodium chloride solution containing 0.125 moles sodium chloride in 500.0 mL of solution.		
Calculate the mass of sodium hydroxide required to make a 0.51 mol L^{-1} solution in a 250 mL volumetric flask.		
What volume of 6 mol L^{-1} HCl is required in the preparation of 100 mL of 1 mol L^{-1} HCl?		
You are supplied with 250 mL of a 0.4 mol L^{-1} solution. You add 250 mL of water to the solution. What is the concentration of the resulting solution?		

5 Put the steps required to determine the number of moles of excess reactant within a mixture into the correct order.

Calculate the number of moles of each reactant.	
Write a balanced chemical equation.	
Subtract the n of excess from the n of limiting.	
Use the molar ratio to determine the limiting factor.	

6 Determine the number of moles of excess reactant in each of the following acid–base reactions.

a A 25 mL sample of 0.17 mol L^{-1} hydrochloric acid is mixed with 50 mL of 0.09 mol L^{-1} sodium hydroxide.

b A 17.2 mL sample of 0.2 mol L^{-1} sulfuric acid is mixed with 32 mL of 0.19 mol L^{-1} potassium hydroxide.

7 Use energy profile diagrams to explain the difference between endothermic and exothermic reactions.

_____

_____

_____

_____

_____

8 A student wanted to determine the molar enthalpy of solution when 4.80 g of potassium nitrate was dissolved in 100.0 mL of water.

a Draw a labelled diagram of the experimental set-up.

b Complete the missing cells in the results table below.

Volume of water (g)	
$m(KNO_3)$ (g)	
Initial temperature (°C)	20.1
Final temperature (°C)	16.7
ΔT (°C)	

c Calculate the value of the molar heat of solution of potassium nitrate. Assume the heat capacity is $4.18 \times 10^3 \, J \, kg^{-1} \, K^{-1}$.

_____

_____

_____

_____

d The theoretical ΔH_{soln} of potassium nitrate is $+34.89\,\text{kJ mol}^{-1}$. Account for the discrepancy between the student's experimental value and the theoretical value.

_____

_____

_____

9 Lead sulfate was dissolved in concentrated sulfuric acid, forming lead hydrogen sulfate.

$$PbSO_4(s) + H_2SO_4(aq) \rightleftharpoons Pb(HSO_4)_2(aq)$$

a What will happen to the concentration of sulfuric acid if lead hydrogen sulfate is removed from the solution?

_____

_____

_____

b How will the equilibrium change if more lead sulfate is added to the solution?

_____

_____

_____

10 Ascorbic acid ionises in water according to the following equation.

$$HC_6H_7O_6 \rightleftharpoons H^+ + C_6H_7O_6^- \quad K_a = 8.0 \times 10^{-5}$$

a Write an equilibrium expression for ascorbic acid.

_____

_____

b Does ascorbic acid ionise to a large or small extent? How do you know?

_____

_____

_____

c A solution is made containing $0.75\,\text{mol L}^{-1}$ ascorbic acid. Calculate the hydrogen ion concentration in this solution.

_____

_____

_____

_____

_____

_____

Properties of acids and bases

WS **5.1** **Using indicators**

STUDENT BOOK
Pages 122, 148, 195

LEARNING GOALS

Interpret pH graphs

Apply the use of indicators

Write a method for the production of a natural indicator

1 Complete the summary table of common indicators used in this course.

Indicator	Colour in pH < 7	pH range of colour change	Colour in pH > 7
Bromothymol blue			
Phenolphthalein			
Methyl orange			
Litmus red			
Litmus blue			

2 A natural indicator was made by boiling beetroot and retaining the juice. The resulting purple solution was tested with a range of substances to determine if they were acidic or basic. The results are recorded below.

Substance	Colour of solution
Lemon juice (acidic)	Purple
Vinegar (acidic)	Purple
Baking soda (basic)	Dark purple
Drain cleaner (basic)	Yellow

a Based on these results, assess the usefulness of beetroot as an acid–base indicator.

_____

_____

_____

_____

_____

_____

b Outline a suitable method to produce another natural indicator, other than beetroot.

_____

_____

_____

_____

_____

_____

c How could the suitability of this natural indicator be tested?

_____

_____

_____

_____

3 The flowers of a hydrangea change colour depending on the pH of the soil. To grow a plant that produces purple flowers, the soil pH must be between 6.5 and 7.

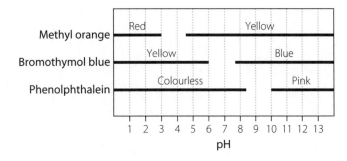

a From the graph above, select the best indicator to determine if the soil pH will produce purple flowers. Justify your choice.

_____

_____

_____

_____

_____

b Before testing the soil with each indicator, barium sulfate powder was added to the surface of the soil sample. Explain why barium sulfate was added.

_____

_____

_____

c Use the information from the graph above to complete the table.

pH	Colour of methyl orange	Colour of bromothymol blue
5.5		
7		
11.5		

d A student used methyl orange and bromothymol blue in order to determine the pH of an unmarked clear solution. The solution turned yellow with both indicators, so the student determined the pH must be 5.5. Assess the validity of the student's claim.

_____

_____

_____

4 List three characteristics that limit the usefulness of indicators.

_____

_____

_____

_____

_____

5 One way to model the reaction of bromothymol blue in an acidic solution is:

$$\text{bromothymol blue} + H_3O^+ \rightleftharpoons \text{bromothymol yellow}$$

a Explain, using Le Chatelier's principle, how the addition of an acid can stress the system of the indicator, and how equilibrium is restored.

_____

_____

_____

_____

_____

b Using an equation similar to the one above, model the reaction of bromothymol blue in an alkali solution.

_____

c Use the equation in part b to explain how the addition of an alkali can stress the system of the indicator, and how equilibrium is restored.

_____

_____

_____

_____

_____

Determine the quantities of neutralising agent required to manage chemical spills and pollutants

Assess the suitability of neutralising agent

Explain the Arrhenius model of bases

Calculate the enthalpy of neutralisation ΔH_{neut}

Write balanced chemical equations

1 To produce electricity, power stations burn fossil fuels such as coal. The burning of fossil fuels produces sulfur dioxide, which is an acidic gas that when mixed with water in the atmosphere produces sulfurous acid (H_2SO_3) falling as acid rain. To prevent this, power stations add quicklime (CaO) at high temperatures to the sulfur dioxide gas, producing calcium sulfite, a solid that can be collected with the ash of the combustion.

a Write a balanced chemical equation for the reaction between sulfur dioxide and quicklime.

_____

b Calculate how many grams of quicklime would be needed to neutralise 140 L of sulfur dioxide at RTP.

_____

_____

2 Hydrochloric acid is a key component in some bathroom cleaners. In transit, hydrochloric acid is often stored alongside solid sodium hydrogen carbonate in case of a spill.

a Write a balanced chemical equation for the reaction between sodium hydrogen carbonate and hydrochloric acid.

_____

b What is the minimum mass of sodium hydrogen carbonate required to neutralise 10.0 L of a $22.0\,\text{mol}\,\text{L}^{-1}$ solution of hydrochloric acid?

_____

_____

_____

3 A chemist has spilled 0.2 kg of solid sodium hydroxide pellets in the lab.

a To neutralise this spill, the chemist poured $6\,\text{mol}\,\text{L}^{-1}$ hydrochloric acid on the floor. What volume of hydrochloric acid would be required?

_____

_____

_____

b Assess the suitability of the chemist using such a concentrated acid to neutralise this spill.

_____

_____

_____

_____

_____

4 In a laboratory, 30.0 mL of a 1.2 mol L^{-1} potassium hydroxide solution was completely neutralised by the addition of 40.0 mL of hydrofluoric acid. It was noted that the temperature of the mixed solution rose from 19.1°C to 24.8°C.

a Write a balanced equation for the reaction between potassium hydroxide and hydrofluoric acid.

___

b Calculate the number of moles of potassium hydroxide used in the neutralisation reaction.

___

___

c Calculate the amount of energy released in the neutralisation reaction.

___

___

d Calculate the number of moles of water produced by the reaction.

___

___

e Calculate the molar enthalpy of neutralisation ΔH_{neut}.

___

___

___

5 A student mixed 4.8 mL of a 0.25 mol L^{-1} solution of sodium hydroxide with a 0.1 mol L^{-1} solution of hydrochloric acid to determine the enthalpy of neutralisation.

The results of the experiment are shown below.

Time (min)	Temperature (°C)		Volume of HCl added (mL)
	NaOH	HCl	
0	21.1	20.9	
1	21.0	21.0	
2	21.0	21.0	
3	21.0	21.0	
4	Solutions are mixed		
5	21.2		3
6	21.5		6
7	21.7		12
8	21.6		15
9	21.5		18
10	21.4		21

a Write a balanced chemical equation for this neutralisation reaction.

___

b Based on recorded results, determine the initial and maximum temperatures ($T_{initial}$ and T_{final}) of this reaction.

___

___

c What volume of HCl was required to completely neutralise the NaOH? Justify your response.

_____

_____

_____

_____

_____

_____

d Calculate the enthalpy of neutralisation ΔH_{neut}.

_____

_____

_____

_____

_____

_____

_____

e The molar heat of neutralisation between a strong acid and a strong base is $-57\,kJ\,mol^{-1}$. Account for possible reasons why the experimental value may differ.

_____

_____

_____

_____

LEARNING GOALS

Describe the contributions of Lavoisier, Davy, Arrhenius and Brønsted–Lowry

Discern the difference between the Arrhenius and Brønsted–Lowry theories of acids and bases

1 'Acids contain hydrogen and can dissolve in water to release hydrogen ions into solution.'
Who originally stated this theory of acids?

A Arrhenius

B Brønsted–Lowry

C Davy

D Lavoisier

2 What contribution did Brønsted–Lowry make to the definition of acids that made it a significant improvement over the Arrhenius model?

A Acids contain hydrogen.

B Acids are proton donors.

C Acids contain oxygen.

D Acids are electron-pair acceptors.

3 In 1887, Svante Arrhenius proposed a definition for acids that was deemed superior to the preceding definitions. Why was Arrhenius' definition seen as a major improvement?

A It explained why some acids do not contain oxygen.

B It showed how the solvent can affect the strength of an acid.

C It showed the relationship between pH and the concentration of H^+ ions.

D It could be used to explain why some acids are strong and others are weak.

4 According to the Arrhenius theory of acids and bases, an acid is a substance that

A tastes sour.

B is capable of donating a hydrogen ion.

C can accept a pair of electrons to form a coordinate covalent bond.

D increases the concentration of hydrogen ions in an aqueous solution.

5 Compare the Arrhenius model of acids and bases to the Brønsted–Lowry model.

	Arrhenius	Brønsted–Lowry
Acid		
Base		
Example chemical equation		

6 Lavoisier, Davy, Arrhenius and Brønsted–Lowry have made significant contributions to the definitions of acids and bases over time. Suggest the limitations of each theory, and how the limitations were addressed by the subsequent theory.

_____

_____

_____

_____

_____

_____

_____

_____

_____

_____

_____

_____

_____

_____

_____

_____

_____

_____

_____

7 Suggest why the Arrhenius' definition is still taught rather than the more sophisticated Brønsted–Lowry definition.

_____

_____

_____

_____

_____

INQUIRY QUESTION: WHAT IS THE ROLE OF WATER IN SOLUTIONS OF ACIDS AND BASES?

WS 6.1 Calculating pH and pOH

STUDENT BOOK
Pages 152–62

LEARNING GOALS

Use data to calculate pH, pOH, [H⁺] and [OH⁻]

Explain the relationship between pH and concentration

Interpret data from a graph

Assess the validity of pH value

1 Draw a flow chart to model the pathway for performing pH calculations for strong acids and bases.

2 Which of the following solutions has the highest pH?

A $0.1\,mol\,L^{-1}\,NH_4Cl$

B $0.1\,mol\,L^{-1}\,CH_3COONa$

C $0.1\,mol\,L^{-1}\,H_2SO_4$

D $0.1\,mol\,L^{-1}\,NaBr$

3 What is the concentration of hydrochloric acid with a pH of 1.74?

A $0.018\,mol\,L^{-1}$

B $55.0\,mol\,L^{-1}$

C $1.8\times10^{12}\,mol\,L^{-1}$

D $5.5\times10^{-13}\,mol\,L^{-1}$

4 The concentration of barium hydroxide is $4.5\times10^{-3}\,mol\,L^{-1}$. Assuming this substance completely disassociated, what is its pH?

A 12

B 11.7

C 2.3

D 2.0

5 The hydrogen ion concentration of acetic acid was $5.27 \times 10^{-5}\,mol\,L^{-1}$. The pH of this solution is correctly expressed as

A 4

B 4.3

C 4.28

D 4.278

6 A solution has a pH of 4.0. The concentration of hydroxide ions in this solution is

A $1.0 \times 10^{-4}\,mol\,L^{-1}$

B $4.0\,mol\,L^{-1}$

C $4.0 \times 10^{-14}\,mol\,L^{-1}$

D $1.0 \times 10^{-10}\,mol\,L^{-1}$

7 Find the values of $[H^+]$, $[OH^-]$ and pOH that correspond with the matching pH values in the table below.

	$[H^+]$ $(mol\,L^{-1})$	pOH	$[OH^-]$ $(mol\,L^{-1})$
pH of orange juice = 3.1			
pH of kombucha = 2.6			
pH of milk = 6.8			
pH of human blood = 7.8			
pH of oven cleaner = 13.2			

8 A student calculated the pH of a $1.2\,mol\,L^{-1}$ acid to be −0.079.

a Assess the validity of the student's calculation.

_____

_____

b How could the student experimentally demonstrate the validity of their calculation?

_____

_____

_____

9 At least one-quarter of the carbon dioxide released from the combustion of fossil fuels dissolves into the world's oceans. When the dissolved carbon dioxide reacts with water, it forms carbonic acid.

a Write a balanced chemical equation for this reaction.

_____

The graph below maps the historical and projected pH and dissolved carbon dioxide in the ocean.

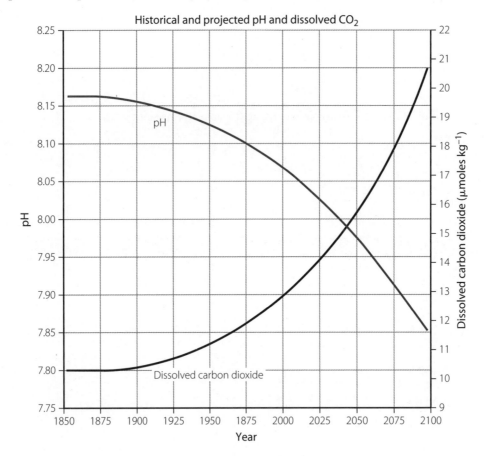

b Based on the recorded pH, determine the [H⁺] in the water for the years 2000 and 2020.

_____

_____

_____

_____

c An ocean sample recorded a hydrogen ion concentration of $8.13 \times 10^{-9}\,mol\,L^{-1}$. In which year was this sample taken?

_____

_____

d How much dissolved carbon dioxide did this sample contain?

_____

e If a sample of sea water contained $1.25 \times 10^{-5}\,mol\,kg^{-1}$ of carbon dioxide, determine the [H⁺].

_____

_____

> **HINT**
>
> 1 micromole $= 1 \times 10^{-6}$ moles

WS 6.2 Comparing concentration and strength of acids

STUDENT BOOK
Pages 152–4

Assess a model used to represent the terms strong, weak, concentrated and dilute

Identify conjugate acid and base pairs

Write the ionisation equations for polyprotic acids

Analyse the validity of controls and equipment

Calculate the ionisation percentage of acids

1 A student drew the following model to illustrate the difference between strong, weak, concentrated and dilute solutions.

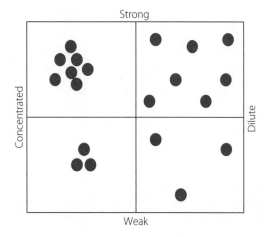

a Assess the usefulness of this model in explaining the key terms.

_____

_____

_____

_____

_____

_____

_____

_____

_____

_____

_____

b Draw and label a correct model to explain the key terms.

2 Label the conjugate pairs in the following equations.

a $NH_3 + H_2O \rightleftharpoons NH_4^+ + OH^-$

b $HI + OH^- \rightleftharpoons I^- + H_2O$

c $HNO_3 + HCO_3^- \rightleftharpoons H_2CO_3 + NO_3^-$

3 **a** Write the corresponding conjugate, identifying it as a conjugate acid or base.

Acid/Base	Conjugate	Acid or base?
H_2O		
OH^-		
HCO_3^-		
H_3PO_4		
H_2S		
HSO_4^-		
ClO_4^-		
$HBrO_2$		

b Identify the three species from the table above that are amphiprotic. Explain why they are amphiprotic.

4 A student is asked to determine the relative strength of nitric acid (HNO_3), formic acid ($HCOOH$) and citric acid ($C_3H_5O(COOH)_3$).

a Write the ionisation reactions for these acids in water.

> **HINT**
>
> Consider the ionisation steps in polyprotic acids.

In order to determine the relative strength of the acids, the student set up three beakers with 100 mL of $0.01 \, mol \, L^{-1}$ of each acid. The student used a pH probe to measure the pH for each acid. The results are shown in the following diagram.

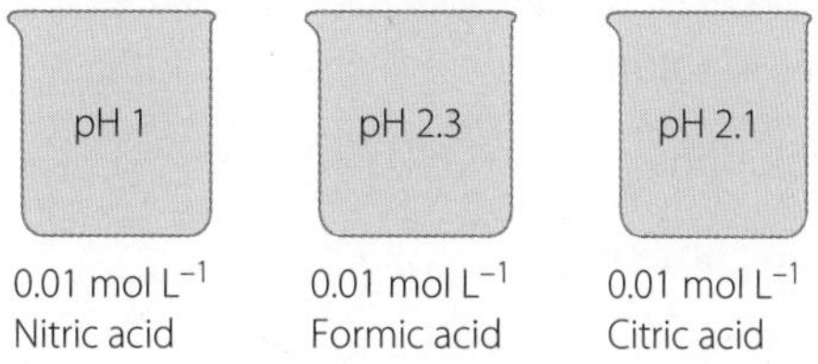

pH 1	pH 2.3	pH 2.1
$0.01 \, mol \, L^{-1}$	$0.01 \, mol \, L^{-1}$	$0.01 \, mol \, L^{-1}$
Nitric acid	Formic acid	Citric acid

b Justify the need for keeping the concentration of each acid constant when determining the relative strength of the acids.

_____

_____

_____

c Explain why a pH probe ensures the *validity* of the measurements.

_____

_____

_____

_____

_____

d Using the pH of the acids from the diagram above, calculate the ionisation percentage of each acid.

_____

_____

_____

e Using all the information supplied, determine the relative strength of the nitric, formic and citric acids.

_____

_____

_____

f 'The extent to which an acid donates protons to water molecules depends on the strength of the conjugate base.' Explain this statement using two acids from this experiment.

_____

_____

_____

_____

_____

_____

Determine the concentration of species in solution

Calculate pH, pOH, [H$^+$] and [OH$^-$] of solutions

Explain the relationship between dilution and pH

Calculate the pH of a solution resulting from mixing acids and bases

Calculate the pH of a solution following dilution

1 A volumetric flask contains 100 mL of 0.01 mol L^{-1} potassium hydroxide solution.

   a A 50 mL aliquot is poured into a second beaker. Calculate the [OH$^-$] in this sample.

   b A 40 mL sample of this aliquot is diluted to 100 mL. Determine the pH of this sample.

   c Explain the effect dilution has on the pH of a basic solution.

   d If the diluted potassium hydroxide solution in part b was further diluted by a factor of 100, what is the new [OH$^-$]?

2 To what volume must 20 mL of a 0.52 mol L^{-1} solution of hydrochloric acid be diluted in order to have a pH of 2.1?

3 A student has 300 mL of sodium hydroxide with a pH of 12.6.

   a What is the [OH$^-$] concentration in this solution?

The student then adds 37 mL of 0.4 mol L^{-1} nitric acid to neutralise the sodium hydroxide.

   b Write a balanced equation for this reaction.

c Determine the minimum volume of nitric acid required to completely neutralise the sodium hydroxide.

d Determine whether the resulting solution will be acidic, basic or neutral. Explain why.

e Determine the pH of the resulting solution.

4 A 1.0 L volumetric flask contains a stock solution of perchloric acid ($HClO_4$) with a pH of 1.4.

Outline a method to prepare a solution of perchloric acid with a concentration of $0.004 \, mol \, L^{-1}$ from this stock. Include all working.

5 A 100 mL sample of a $0.015 \, mol \, L^{-1}$ solution of hydrochloric acid was added to a beaker containing 100 mL of a $0.05 \, mol \, L^{-1}$ barium hydroxide solution.

a Calculate the pH of the resulting solution.

b The solution from part a was then diluted to a final volume of 350 mL. Calculate the pH of this dilution.

_____

_____

_____

6 A 1.6 g sample of sodium hydroxide was dissolved into 50 mL of water. It was then mixed with 150 mL of $0.28\,mol\,L^{-1}$ sulfuric acid solution. Calculate the final pH of the solution.

_____

_____

_____

_____

_____

_____

_____

_____

_____

_____

_____

7 Use the pH indicator chart below to answer the following questions.

Red cabbage indicator chart

Colour	red		violet		purple		blue		green		yellow		
pH	1	2	3	4	5	6	7	8	9	10	11	12	13	14

a What colour would the red cabbage indicator turn when added to a $0.012\,mol\,L^{-1}$ solution of hydrobromic acid?

_____

_____

b What colour will the solution turn if 8.0 mL of the solution from part a was diluted to 100 mL?

_____

_____

c What volume, in mL, of $0.0040\,mol\,L^{-1}$ potassium hydroxide solution will be required to neutralise 30 mL of the diluted solution from part b?

_____

_____

_____

_____

_____

STUDENT BOOK
Pages 165–72

WS 6.4 Using dissociation constants

LEARNING GOALS

Use K_a values to determine pH

Describe the relationship between [H⁺] and K_a

Draw a graph and extrapolate data

Explain the relationship between K_w and temperature

Use pK_a values to determine pH

1 a Write the expression for the acid dissociation constant of acetic acid.

b If the K_a of acetic acid is 1.8×10^{-5}, calculate the pH of $0.30\,\text{mol L}^{-1}$ solution of acetic acid.

2 The table below shows the K_w values and pH of pure water at increasing temperatures.

T (°C)	K_w	pH
0	0.1×10^{-14}	7.47
10	0.3×10^{-14}	7.27
20	0.7×10^{-14}	7.08
25	1.0×10^{-14}	7.00
30	1.5×10^{-14}	
40	2.9×10^{-14}	6.77
50	5.5×10^{-14}	6.63
100	51.3×10^{-14}	6.14

a Draw a graph to show the relationship between the temperature and pH of pure water.

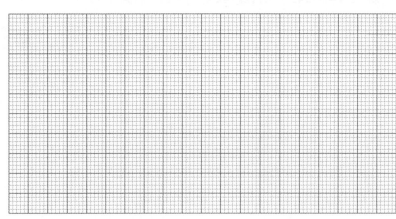

b Use your graph to determine the pH at 30°C.

_____

c Explain the relationship between K_w and temperature.

_____

_____

_____

3 Acetic acid and formic acid are both weak acids. Their K_a values at 25°C are:

acetic acid (CH$_3$COOH): $K_a = 1.8 \times 10^{-5}$

formic acid (HCOOH): $K_a = 1.77 \times 10^{-4}$

a Write an equation for the reaction of formic acid and water.

_____

b Write the K_a expression for formic acid.

_____

c Calculate the pK_a value for formic acid.

_____

d Calculate the pH of a solution of 0.250 mol L^{-1} formic acid.

_____

_____

_____

_____

_____

_____

_____

_____

e Based on K_a values provided, which acid – acetic or formic – would have a higher pH at the same temperature and concentration?

_____

_____

4 The data below was recorded for two unknown acids at 25°C.

HX	$pK_a = 3.86$
HY	$pK_a = 1.92$

a Which acid, HX or HY, is stronger? Explain your answer.

_____

_____

_____

_____

b Using the formula $K_w = (K_a)(K_b)$, demonstrate why HX produces the strongest conjugate base.

_____

_____

_____

_____

c Determine the $[H^+]$ concentration and pH in $0.230 \, \text{mol L}^{-1}$ of HY.

_____

_____

_____

_____

_____

_____

_____

_____

_____

_____

7 Quantitative analysis

WS 7.1 Analysing titrations

STUDENT BOOK
Pages 190–95

LEARNING GOALS

Identify errors in procedures for conducting titrations

Evaluate the effect of incorrect procedures

Process titration data to calculate the concentration of unknown solutions

Justify the choice of indicators

Propose appropriate safety procedures

Explain how to minimise identified systematic errors

1 Two students needed to prepare a standard solution of sodium carbonate. They weighed the required amount of Na_2CO_3 on a watch glass on an electronic balance. They filled a 500 mL volumetric flask with tap water so that the top of the meniscus was sitting exactly on the mark. They then transferred the Na_2CO_3 to the flask, stoppered the flask and shook it well to ensure uniform mixing.

a Identify all the errors in the students' procedure.

_____

_____

_____

_____

b Describe how the students should have correctly prepared the standard solution.

_____

_____

_____

_____

_____

_____

_____

_____

c Explain the effect, if any, of these errors on the calculated concentration compared to the actual concentration of the solution produced.

_____

_____

_____

_____

_____

_____

_____

_____

2 Potassium hydrogen phthalate, $KH(C_8H_4O_4)$ is ideal to make a primary standard solution.

a Provide two properties that would make potassium hydrogen phthalate suitable for making a primary standard and explain why these properties are important.

_____

_____

_____

b Calculate the mass required to make 150 mL of a 0.020 $mol\,L^{-1}$ solution of potassium hydrogen phthalate ($KH(C_8H_4O_4)$).

_____

_____

_____

3 A volume of a standard sodium carbonate solution was pipetted into a conical flask and a few drops of phenolphthalein indicator were added. A burette was filled with an HCl solution and the sodium carbonate solution was titrated with the acid. This procedure was repeated three times.

a Explain how the pipette should be prepared for use. Justify your response.

_____

_____

_____

_____

_____

b Explain how the burette should be prepared for use. Justify your response.

_____

_____

_____

_____

_____

4 The electrolyte in car batteries is sulfuric acid. A curious student decided to determine the concentration of this acid in a well-charged car battery. They took exactly 2.00 mL of the battery acid using a pipette and titrated it with a 1.16 mol L^{-1} sodium hydroxide solution. They found 17.1 mL was needed to reach the equivalence point.

a Describe what safety procedures the student should have followed.

_____

_____

_____

_____

_____

b Calculate the molarity of the sulfuric acid in the battery.

_____

_____

_____

_____

c The battery manufacturer advertised the concentration of acid as between 4.2 and 4.8 mol L^{-1}. Suggest how the student could have improved the accuracy and reliability of their results.

_____

_____

_____

_____

_____

_____

5 A sample of lemon juice was analysed by a student. They took 25.00 mL of the juice and diluted it to 200 mL. Exactly 25.00 mL of the diluted juice was titrated against 0.1045 mol L^{-1} sodium hydroxide. In each case, the same indicator colour was obtained at the end point.

a Suggest an appropriate indicator to use in this titration and identify the colour change that should be observed. Justify your choice.

_____

_____

The student's results are recorded below.

Attempt	mL NaOH added
1	25.70
2	23.95
3	24.15
4	24.10
Average	

b Calculate the average titre of NaOH required. Write your answer in the table above.

c Suggest why the first attempt yielded a higher result.

_____

_____

d **i** Through research the student found that lemon juice contains citric acid, which has the formula ($C_6H_8O_7$) and is triprotic. Write a balanced chemical equation to represent the reaction between citric acid and sodium hydroxide.

_____

ii Calculate the concentration of citric acid in the undiluted lemon juice.

_____

_____

_____

_____

_____

e Explain what assumption(s) have been made in the calculation in part d ii and how these may affect the accuracy of the calculated value.

_____

_____

_____

_____

6 A chemist was asked to determine the concentration of ammonia in a 1-litre bottle of heavy-duty floor cleaner claiming to contain cloudy ammonia. The chemist pipetted 20.0 mL of the cloudy ammonia solution into a 100 mL conical flask.

A 70.0 mL sample of 0.100 mol L^{-1} hydrochloric acid was immediately added to the conical flask and reacted with the ammonia. The excess hydrochloric acid was titrated against 0.0250 mol L^{-1} sodium carbonate solution and required 12.5 mL on average.

Calculate the concentration of the ammonia in the cloudy ammonia solution.

_____

_____

_____

_____

_____

_____

_____

_____

_____

7 Kiwi fruit are high in ascorbic acid, $HC_6H_7O_6$, a weak diprotic acid.

Fourteen pieces of kiwi fruit were processed to give 80 mL of juice. A 20 mL sample of this juice was diluted to 100 mL in a volumetric flask to ensure a clear solution. The diluted kiwi juice was titrated against 0.100 mol L^{-1} sodium hydroxide solution that had been standardised with oxalic acid dihydrate, $(COOH)_2.2H_2O$.

a Outline the method used to standardise the sodium hydroxide solution.

_____

_____

_____

_____

_____

_____

_____

_____

_____

_____

_____

_____

_____

b Identify three steps needed to determine the concentration of ascorbic acid in the kiwi fruit and justify how these steps ensure valid and reliable results.

Step	Justification

Systematic errors occur when measurements are shifted from their true value. Although they do not affect the reliability of results, they do affect the accuracy.

c Identify one possible source of systematic error within your method.

_____

_____

d Outline how the error identified in part c may affect the results.

_____

_____

_____

e Explain how the chance of this error occurring could be reduced.

_____

_____

The student obtained the following results with 25.00 mL aliquots of diluted kiwi fruit solution titrated against standardised $0.100 \, mol \, L^{-1}$ sodium hydroxide.

Attempt	mL NaOH added
1	13.60
2	11.63
3	10.48
4	11.71
5	11.64

f Calculate the average titre, identifying which particular values were used and why.

_____

_____

_____

g Calculate the concentration of ascorbic acid in a kiwi fruit.

_____

_____

_____

_____

_____

_____

_____

LEARNING GOALS

Solve problems by interpreting titration curves

Use data to draw a titration curve

Calculate the concentration of acid used using data from a titration curve

Explain the reason for the shape of polyprotic titration curves

Identify the equivalence point of conductivity curves

Explain the reason for the shape of conductivity curves

1 The titration curves below were obtained by titrating equal volumes of two different acids with the sodium hydroxide. What conclusions can be drawn regarding the strength and concentration of these acids?

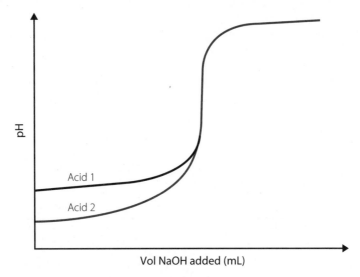

A Same concentration but Acid 1 is stronger than Acid 2

B Same concentration but Acid 1 is weaker than Acid 2

C Same strength but Acid 1 is more concentrated than Acid 2

D Same strength but Acid 1 is less concentrated than Acid 2

2 For which titration would the use of methyl orange be a suitable indicator?

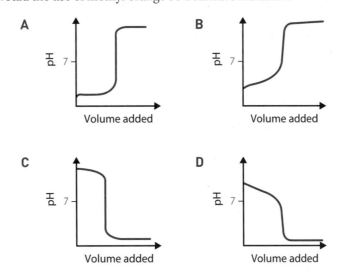

3 The diagram below is the titration curve for the neutralisation of an acid with a base.

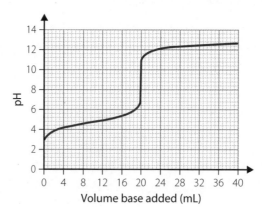

What pH range should an indicator have in this titration?

A 3.1–4.4

B 5.0–8.0

C 6.0–7.6

D 8.3–10.0

4 Which of the following conductivity graphs represents the titration of H_2SO_4 with ammonium hydroxide, a weak base?

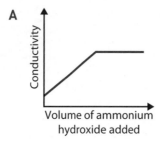

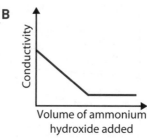

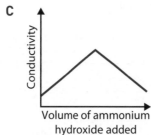

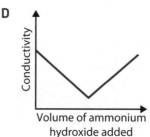

5 Which acid–base combination best reflects this curve?

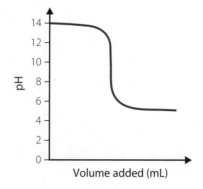

A $HF(aq) + NH_3(aq) \rightarrow NH_4F(aq)$

B $HCl(aq) + NaOH(aq) \rightarrow NaCl(aq) + H_2O(l)$

C $CH_3COOH(aq) + NaOH(aq) \rightarrow CH_3COONa(aq) + H_2O(l)$

D $H_2SO_4(aq) + 2NH_3(aq) \rightarrow (NH_4)_2SO_4(aq)$

6 A student recorded the pH of an acid–base titration using a pH probe. Their results are shown below.

pH	Volume of 0.1 mol L^{-1} base added (mL)
3.0	0
4.2	5
4.8	10
5.0	15
5.6	20
8.6	25
12.2	30
12.6	35
12.8	40
12.8	45
12.8	50

a Draw the titration curve of the results.

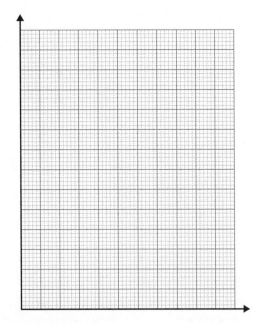

b Label the equivalence point on your titration curve.

c Identify a suitable acid and base combination for this curve. Justify your response.

_____

_____

_____

d What is the volume of base required to completely neutralise the weak acid?

_____

e Assuming the acid and base were both monoprotic, calculate the initial concentration of acid used. The titration was performed with a 50.0 mL aliquot of acid.

_____

_____

_____

_____

7 Explain the reason for the shape of the titration curve below. Use chemical equations to support your answer.

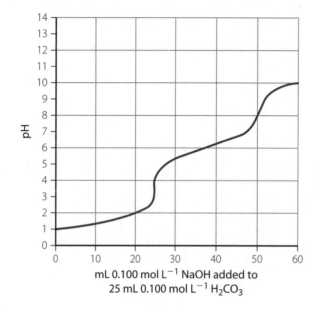

mL 0.100 mol L^{-1} NaOH added to
25 mL 0.100 mol L^{-1} H$_2$CO$_3$

_____

_____

_____

_____

8 A student recorded the following data in a conductometric titration of a strong acid with a weak base.

Conductance (ms cm^{-1})	Volume of weak base added (mL)
6.2	0
5.8	5
4.6	10
3.8	15
3.0	20
3.0	25
3.0	30
3.0	35

a Graph the conductometric titration curve of this data.

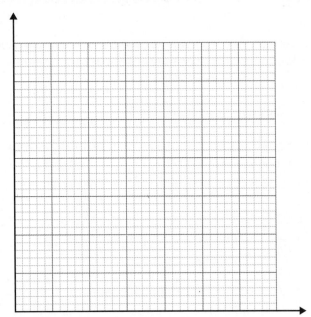

b Label the equivalence point on the curve.

c Explain the changes in conductance that occur.

_____

_____

_____

_____

_____

_____

Determine the percentage of calcium carbonate in limestone through back titration

Analyse experimental procedure

Evaluate the steps in determining the protein content of cereals

1 A student conducted an investigation to determine the percentage of calcium carbonate in a sample of limestone. They decided to use a back titration for the investigation.

a Explain why the student chose a back titration rather than a direct titration.

_____

_____

_____

_____

_____

The student dissolved a 5.0 g sample of the limestone in 100 mL of 0.85 mol L^{-1} HCl. The solution was heated to ensure that all the carbon dioxide was removed, then cooled and diluted to 250 mL.

b **i** Explain what the student needed to ensure about the amount of HCl added.

_____

_____

_____

_____

ii How could they check there was excess HCl?

_____

_____

iii Explain why they heated the solution and possible consequences of not heating.

_____

_____

_____

_____

_____

_____

_____

_____

A 25.0 mL sample of the diluted solution was titrated with 26.65 mL of a standardised solution of $0.15 \, mol \, L^{-1}$ sodium hydroxide.

c **i** Explain why sodium hydroxide should be standardised before being used in this titration.

_____

_____

_____

_____

_____

ii Explain what the best indicator would be to use in this titration and why.

_____

_____

_____

d Determine the percentage of calcium carbonate in the original limestone sample.

_____

_____

_____

_____

_____

_____

_____

_____

_____

_____

_____

_____

_____

_____

_____

_____

2 The protein found in breakfast cereal can be oxidised in the presence of sulfuric acid to form the ammonium ion. The ammonium ion can be converted into ammonia and removed by distillation using hydrochloric acid. The amount of unreacted acid is determined by back titration.

The following method was used to determine the protein content in a breakfast cereal.

1 Transfer 2.0 g of dry and finely ground cereal into a conical flask.

2 Add 0.7 g of a HgO catalyst, 10 g of K_2SO_4 and 25 mL of concentrated H_2SO_4.

3 Bring the solution to a boil. Continue boiling until the solution turns clear and then boil for at least an additional 30 minutes. Cool to room temperature.

4 Remove the Hg^{2+} catalyst by adding 200 mL of H_2O and 25 mL of 4% w/v K_2S.

5 Pour the solution into a distillation apparatus, adding 25 g of HCl and marble boiling chips.

6 Distil the NH_3 into a collection flask containing a known amount of standardised HCl.

7 Titrate the excess HCl with a standard solution of NaOH using bromothymol blue as an indicator.

a Match the chemical process to the steps in which it occurs within the method.

Chemical process	Procedure step
The protein found in cereal can be oxidised to form the ammonium ion.	
The ammonium ion can be converted into ammonia.	
The amount of unreacted acid is determined by back titration.	

b Oxidising the protein converts all the nitrogen into an ammonium ion. Why is the amount of nitrogen not determined by titrating the ammonium ion with a strong base?

_____

_____

_____

_____

_____

c Ammonia is highly volatile. How may this lead to an error within this method?

_____

_____

_____

WS 7.4 **Applying buffers**

LEARNING GOALS

Describe the main features of buffers

Explain buffering capacity

Write balanced equations of buffer solutions

Explain the importance of a named buffer on a natural system

1 Complete each of the following statements.

   a A buffer is a solution that resists change in pH because it contains comparable amounts of a weak acid and

      its _____ .

   b The natural buffering that occurs in some lakes is a result of the dissolution of _____ from the air.

   c The other half of a buffer system containing sodium acetate would be _____ .
   Carbon dioxide is one component in the complex buffer system in blood.

   d The CO_2 dissolves in blood (water) to produce _____ .

   e The other part of the buffer system is the _____ ion.

   f If the concentration of OH^- ions increase in the blood, H_2CO_3 reacts to _____ the OH^- concentration.

2 Explain what is meant by the term 'buffering capacity'.

   _____

   _____

   _____

   _____

3 Often buffers are created with equal concentrations of acid and conjugate base. Explain why this is the case.

   _____

   _____

   _____

4 Dihydrogen phosphate and monohydrogen phosphate ions play an important role in maintaining the pH in intracellular fluid. Write an equation to show this buffer system.

   _____

5 Explain how the conjugate acid–base pair $H_2PO_4^-$ / HPO_2^{2-} can act as a buffer system. Include a chemical equation in your answer.

   _____

   _____

   _____

   _____

   _____

   _____

_____

_____

_____

_____

_____

_____

6 A buffer solution was created using equal volumes and concentrations of hydrofluoric acid and sodium fluoride. Explain how the pH of this solution would be affected by the addition of a small amount of sodium hydroxide solution. Include an equation in your answer.

_____

_____

_____

_____

_____

_____

7 A buffer is made by mixing 10 mL of 0.40 mol L^{-1} sodium dihydrogen phosphate with 10 mL of 0.40 mol L^{-1} sodium hydrogen phosphate.

The three K_a values for phosphoric acid are: $K_{a1} = 7.5 \times 10^{-3}$, $K_{a2} = 6.2 \times 10^{-8}$, $K_{a3} = 4.8 \times 10^{-13}$.

a Identity the conjugate acid and base pair in this reaction.

_____

b Based on this acid base pair, identify the most appropriate K_a value to use in the calculation. Justify your choice.

_____

_____

_____

c Calculate the pH of this solution.

_____

_____

Module six: Checking understanding

1 Using the K_a, which of the following is the strongest acid?

   A Nitrous acid ($K_a = 4.0 \times 10^{-4}$)

   B Benzoic acid ($K_a = 6.46 \times 10^{-5}$)

   C Lactic acid ($K_a = 1.38 \times 10^{-4}$)

   D Hydrocyanic acid ($K_a = 6.17 \times 10^{-10}$)

2 Which beaker contains a dilute strong acid?

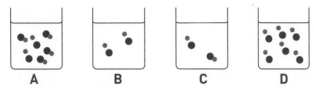

3 The pK_a of carbonic acid is 6.37. What is the dissociation constant (K_a)?

   A 2.34×10^5

   B 8.04×10^{-1}

   C 6.37×10^{-10}

   D 4.27×10^{-7}

4 Which of the following is an Arrhenius base?

   A Ca

   B $CaCO_3$

   C $Ca(OH)_2$

   D $Ca(HCO_3)_2$

5 Consider the following equilibrium:

$$HF(aq) + CF_3COO^-(aq) \rightleftharpoons F^-(aq) + CF_3COOH(aq) \qquad K_{eq} = 3.80 \times 10^{-4}$$

Which row of the table correctly identifies the acid in highest concentration, and its conjugate base in this system?

	Acid	Conjugate base
A	CF_3COOH	CF_3COO^-
B	CF_3COOH	HF
C	HF	F^-
D	HF	CF_3COO^-

6 A 25.0 mL sample of a $0.100 \, mol \, L^{-1}$ acid is titrated against a sodium hydroxide solution. Which of the following acids would require the greatest volume of sodium hydroxide to reach the end point?

   A Citric acid

   B Sulfuric acid

   C Nitric acid

   D Hydrochloric acid

7 A solution of hydrochloric acid was found to have a hydrogen ion concentration of $7.21 \times 10^{-2} \, mol \, L^{-1}$. What is the pH of this acid?

   A 1.142

   B 1.14

   C 1.1

   D −1.1

8 One litre of an aqueous solution is formed from mixing equal volumes of $0.3\,\text{mol}\,\text{L}^{-1}$ sulfuric acid and $0.3\,\text{mol}\,\text{L}^{-1}$ magnesium sulfate. As a buffer, how effective is this solution?

A Ineffective, because sulfuric acid is a strong acid

B Ineffective, because magnesium sulfate forms an acidic salt

C Effective, because SO_4^{2-} is the conjugate base of H_2SO_4

D Effective, because the pH would change when a solution of NaOH is added

9 The pH of a $0.080\,\text{mol}\,\text{L}^{-1}$ solution of acetic acid is 2.9.

What percentage of the acetic acid has dissociated into ions?

A 1.1%

B 1.3%

C 1.6%

D 2.8%

10 Which of the following substances is an amphiprotic salt?

A Sodium hydrogen carbonate

B Sodium chloride

C Sodium sulfate

D Sodium iodide

11 According to Davy and Arrhenius, all acids contain hydrogen atoms. However, there are many compounds that contain hydrogen atoms and are not classified as acids.

a Explain this statement using the Brønsted–Lowry theory.

_____

_____

_____

_____

b The following compounds all contain hydrogen atoms. Explain their classification as an acid according to Arrhenius and Brønsted–Lowry.

Compound	Arrhenius	Brønsted–Lowry
Ethane (C_2H_6)		
Ethanoic acid (CH_3COOH)		
Ammonia (NH_3)		

12 A student adds 50.0 mL of a 0.124 mol L^{-1} sulfuric solution to 60.0 mL of a 0.200 mol L^{-1} lithium hydroxide solution.

    a Determine if the resulting solution would be acidic, neutral or basic. Explain your response.

    b Calculate the pH of the resulting solution.

The student wants to confirm their calculation by testing the pH of the solution using a natural indicator.

    c Outline a method that could be used to make a natural indicator. Include expected results.

    d Suggest how the accuracy of this investigation could be improved.

13 A solution of sodium hydroxide was titrated against a standardised solution of acetic acid, which had a concentration of 0.365 mol L^{-1}. End point was reached when 12.5 mL of sodium hydroxide solution had been added to 20.0 mL of the acetic acid solution, turning the phenolphthalein indicator a faint pink colour.

    a Calculate the concentration of the sodium hydroxide solution.

b Identify the analyte and the titrant.

_____

_____

c Justify at least **four** named steps this student could have performed in their replication of this titration to reduce experimental error and ensure their results were valid and reliable.

Step	Justification

14 A conductometric titration was undertaken to determine the concentration of a potassium hydroxide solution. The potassium hydroxide solution was added to 200.0 mL of standardised 2.215×10^{-4} mol L^{-1} sulfuric acid solution. The results of the titration are shown in the conductivity graph.

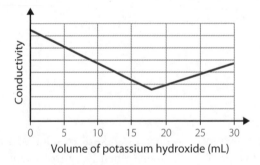

a Label the equivalence point on the curve.

b Calculate the concentration of the potassium hydroxide.

_____

_____

_____

_____

_____

_____

c Explain why the conductivity increases after the equivalence point is reached, and why it does not match the gradient of the slope before equivalence is reached.

_____

_____

_____

_____

15 A buffer was prepared with methanoic acid and sodium methanoate. A few drops of universal indicator were then added. When $2\,\text{mL}$ of $0.1\,\text{mol}\,\text{L}^{-1}$ HCl(aq) or $2\,\text{mL}$ of $0.1\,\text{mol}\,\text{L}^{-1}$ NaOH(aq) was added there were no changes in the colour of the solution.

a Explain these observations. Support your answer with at least one chemical equation.

_____

_____

_____

_____

_____

_____

_____

_____

b Write the expression for the acid dissociation constant of methanoic acid.

_____

_____

c If the K_a of methanoic acid is 1.77×10^{-4}, calculate the pH of a $0.30\,\text{mol}\,\text{L}^{-1}$ solution of methanoic acid.

_____

_____

_____

_____

_____

_____

_____

Reviewing prior knowledge

1 For each statement, write true or false. For false statements, rewrite them so they are true.

   a Valency is the charge on an element.

   _____

   b Carbon has a valency of 6.

   _____

   c Hydrogen and halogens have a valency of one.

   _____

   d Oxygen has a valency of 8.

   _____

   e Nitrogen has a valency of 5.

   _____

   f Combustion of hydrocarbons always produces water.

   _____

   g Combustion reactions are endothermic.

   _____

   h Hydrocarbons contain carbon, hydrogen and oxygen.

   _____

   i The formula for glucose is $C_{12}H_{22}O_{11}$.

   _____

   j Acetic/ethanoic acid is an organic acid.

   _____

   k Fractional distillation is used to separate components of a mixture based on boiling point.

   _____

   l Crude oil is a pure substance.

   _____

   m Nylon is an example of a polymer.

   _____

   n Soaps can be made from naturally occurring substances.

   _____

   o Oxidation is loss of electron(s) while reduction is gain in electron(s).

   _____

2 Draw the structural formula for each of the following.

   a Methane

   b Ethanol

   c Acetic/ethanoic acid

3 Show and label the intermolecular forces between the following molecules.

   a Two methane molecules

   b Three methanol molecules

4 **a** Write equations to show the complete combustion of the following compounds.

     i Octane, $C_8H_{18}(l)$

     ii Ethanol, $C_2H_5OH(l)$

b Which compound in part a above is more likely to undergo incomplete combustion? Explain your answer.

_____

_____

_____

_____

c A student noticed black soot forming at the bottom of the beaker when she placed her lit Bunsen burner under it.

   i State the likely flame colour of the Bunsen burner that caused this.

_____

   ii Explain the observation described above including an equation as part of your answer. Assume that methane, CH_4, was the gas used for the Bunsen burner.

_____

_____

5 Acidified potassium dichromate, $K_2Cr_2O_7$, was reacted with a substance Y(s) to produce Y^{2+}(aq) and chromium ions, Cr^{3+}(aq).

  a Write the oxidation half-equation.

_____

  b Write the reduction half-equation.

_____

  c Write the NET equation.

_____

  d Identify the following in the above net equation.

    Oxidant: _____

    Reductant: _____

6 Calculate the mass of ethanol, C_2H_5OH, that must be burnt in a spirit burner to increase the temperature of 155 g water by 35.0°C, assuming no heat is lost to the environment. The molar heat of combustion of ethanol is $1367\,kJ\,mol^{-1}$.

_____

_____

_____

_____

_____

_____

8 Nomenclature

WS 8.1 Investigating IUPAC nomenclature of hydrocarbons and haloalkanes

STUDENT BOOK
Pages 239–54

LEARNING GOALS

Identify alkanes, alkenes, alkynes and haloalkanes from given formula

Identify saturated and unsaturated hydrocarbons

Name and draw structural formulae of alkanes, alkenes and alkynes

Distinguish between alicyclic, aliphatic and aromatic hydrocarbons

1 Use the words from the list to fill in the gaps. A word may be used more than once.

alicyclic	carbon	double	triple
aliphatic	cyclohexene	hydrogen	unsaturated
aromatic	cyclopropane	single	

Hydrocarbons contain only the elements _____ and _____. When compounds

form from these elements with straight or branched chains, they are referred to as _____

hydrocarbons. These compounds can be saturated, i.e. they have only _____ bonds or can be

unsaturated, which means they have _____ or _____ bonds. When hydrocarbons

form ring structures, they are referred to as _____ hydrocarbons. The simplest member of

this series is _____, which has a molecular formula of C_3H_6. These compounds can be saturated

or _____. An example of such an _____ compound that is used in school laboratories

is _____, C_6H_{10}. Benzene, C_6H_6, belongs to a group called _____ hydrocarbons.

2 Complete the table to identify each substance as an alkane, alkene or alkyne and saturated or unsaturated.

Condensed formula	Alkane/alkene/alkyne	Saturated/unsaturated
CH_3CHCH_2		
$CHCCH_2CH_3$		
CH_3CH_3		
$CH_2CHCH_2CH_3$		
$CH_3CH_2CH_3$		
CH_3CCCH_3		

3 State the IUPAC names of the following compounds.

Compound	IUPAC Name
a $$H-\underset{\underset{H}{\mid}}{\overset{\overset{H}{\mid}}{C}}-\underset{\underset{H}{\mid}}{\overset{\overset{H}{\mid}}{C}}-\underset{\underset{H}{\mid}}{\overset{\overset{H}{\mid}}{C}}-\underset{\underset{H}{\mid}}{\overset{\overset{H}{\mid}}{C}}-H$$	
b $$H-\underset{\underset{H}{\mid}}{\overset{\overset{H}{\mid}}{C}}-\underset{\underset{\underset{H-C-H}{\mid}}{\mid}}{\overset{\overset{H}{\mid}}{C}}-\underset{\underset{H}{\mid}}{\overset{\overset{H}{\mid}}{C}}-\underset{\underset{H}{\mid}}{\overset{\overset{H}{\mid}}{C}}-H$$	
c $H_3C-CH_2-CH-CH-CH_2-CH_3$ with CH_3 and CH_2-CH_3 branches (CH_2-CH_3)	
d $H_3C-CH-CH_2-CH-CH_3$ with I and Br branches	
e $$H-\underset{\underset{H}{\mid}}{\overset{\overset{H}{\mid}}{C}}-\overset{\overset{H}{\mid}}{C}=\overset{}{C}-\underset{\underset{H}{\mid}}{\overset{\overset{H}{\mid}}{C}}-H$$	
f $H_3C-CH-CH-CH_2-CH-CH_3$ with CH_2-CH_3, CH_3 and CH_2-CH_3 branches	
g $CH=C-C-CH_3$ with CH_3, CH_3, CH_3 and H branches	
h $$F-\underset{\underset{F}{\mid}}{\overset{\overset{F}{\mid}}{C}}-\underset{\underset{Cl}{\mid}}{\overset{\overset{Cl}{\mid}}{C}}-Cl$$	

4 Draw the structural formulae of the compounds named below.

	IUPAC Name	Structural formula
a	2,2-Dimethylpropane	
b	Methylpropene	
c	3-Ethyl-3-methylheptane	
d	2,3-Dichloropent-2-ene	
e	Tetrachloromethane	
f	3-Fluoro-1,1,2-tribromopent-1-ene	
g	1,3,3-Tribromo-4-chlorobut-1-yne	
h	3,4,4,5-Tetramethylheptane	

WS 8.2

Investigating IUPAC nomenclature of functional groups

STUDENT BOOK
Pages 239–51

LEARNING GOALS

- Identify functional groups from given formula
- Draw structural formulae of alcohols
- Write condensed formula for functional groups
- Apply naming priority when multiple functional groups are present
- Apply IUPAC nomenclature to organic compounds

1 Fill in the table to complete the missing cells.

	Structural formula	Condensed formula	IUPAC name
a	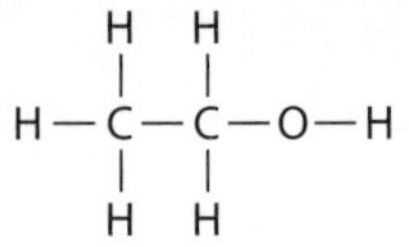		
b		$(CH_3)_3COH$	
c			Propanal
d	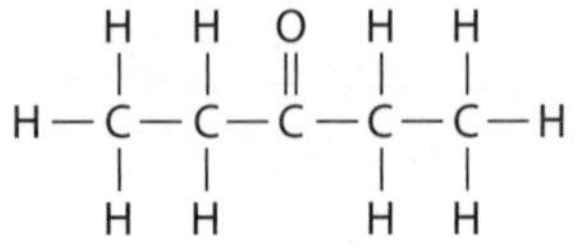		
e		HCOOH	
f			Ethanamine
g	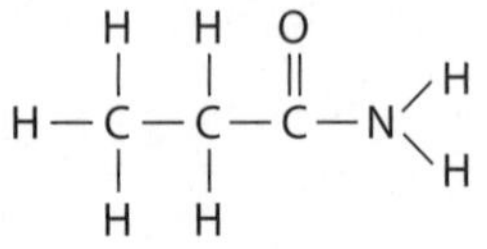		

2 State the IUPAC name for the compounds shown in the table.

	Compound	IUPAC name
a	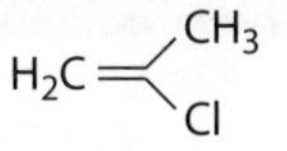	
b	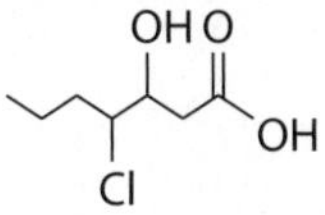	
c	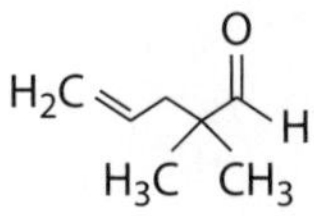	
d	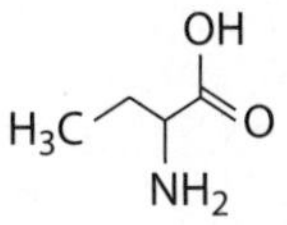	
e	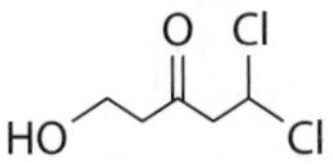	
f	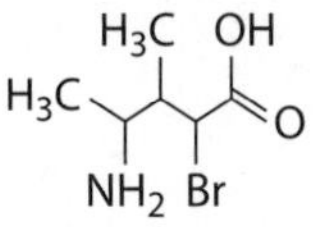	
g	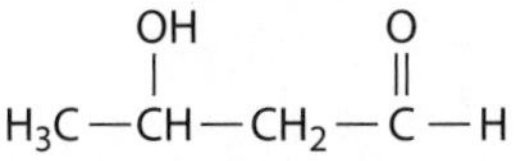	
h	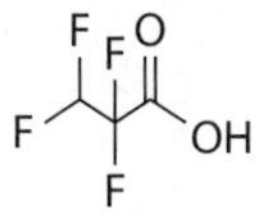	

1 Draw and name structural isomers of C_5H_{12}.

2 Draw and name five structural isomers of C_6H_{12}.

3 a Draw and name the isomers with molecular formula $C_4H_{10}O$.

   b Identify the isomers in part a that could be classified as:

    i position isomers

_____

_____

    ii chain isomers.

_____

_____

4 Draw and name the functional group isomers with molecular formula C_3H_6O.

5 Describe how the isomers shown could be classified.

Isomers		Chain/position/functional group
a	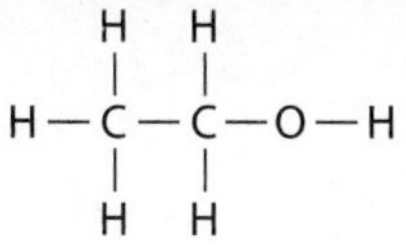	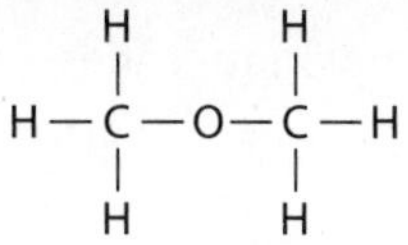
b	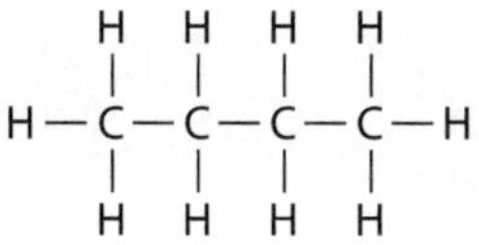	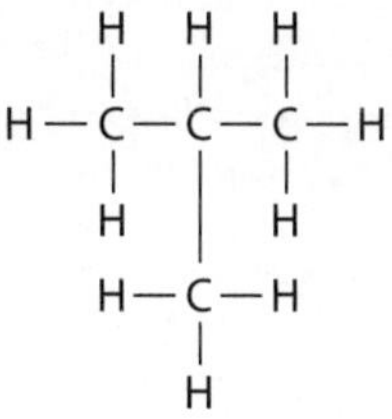
c		
d	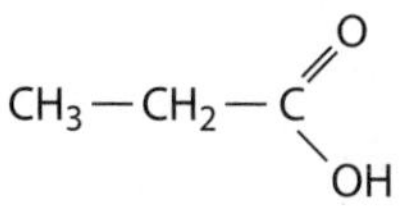	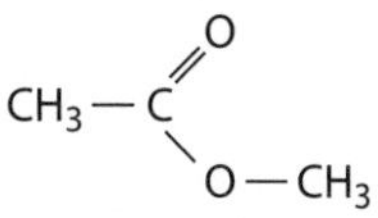
e	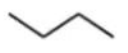	
f	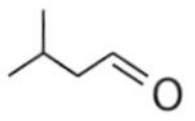	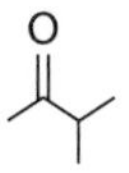
g	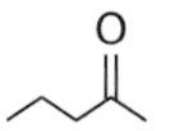	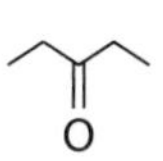

6 Identify which groups of organic compounds cannot have position isomers.

_____

_____

9 Hydrocarbons

STUDENT BOOK
Pages 256–7; 263–4

INQUIRY QUESTION: HOW CAN HYDROCARBONS BE CLASSIFIED BASED ON THEIR STRUCTURE AND REACTIVITY?

WS 9.1 Investigating properties within a homologous series

LEARNING GOALS

Design methods to distinguish between compounds

Compare properties of compounds within and between homologous series

Identify and explain trends

1 The table shows the boiling points of some alkanes, alkenes and alkynes.

Alkane	Formula	Boiling point (°C)	Alkene	Formula	Boiling point (°C)	Alkyne	Formula	Boiling point (°C)
Methane	CH_4	−163						
Ethane	C_2H_6	−88	Ethene	C_2H_4	−104	Ethyne	C_2H_2	−84
Propane	C_3H_8	−42	Propene	C_3H_6	−47	Propyne	C_3H_4	−48
Butane	C_4H_{10}	−0.5	But-1-ene	C_4H_8	−6	But-1-yne	C_4H_6	8
Pentane	C_5H_{12}	36	Pent-1-ene	C_5H_{10}	30	Pent-1-yne	C_5H_8	40
Hexane	C_6H_{14}	69	Hex-1-ene	C_6H_{12}	64	Hex-1-yne	C_6H_{10}	71
Heptane	C_7H_{16}	98	Hept-1-ene	C_7H_{14}	93	Hept-1-yne	C_7H_{12}	100
Octane	C_8H_{18}	125	Oct-1-ene	C_8H_{16}	122	Oct-1-yne	C_8H_{14}	127

a Draw a graph for the boiling points for each of the homologous series against the number of carbons on the grid below.

b Explain the trends in boiling points shown in the graph.

_____

_____

_____

_____

_____

_____

_____

_____

_____

c Pentane and 2,2-dimethylpropane are isomers. Draw the structural formula for each compound and explain whether you expect the two compounds to have the same boiling point.

_____

_____

_____

_____

_____

d A chemist needs to separate a 100 mL mixture of hexane, hex-1-ene and hex-1-yne to obtain the pure components. Outline a safe method to separate the components of the mixture.

_____

_____

_____

_____

_____

_____

_____

_____

2 The labels have fallen off two bottles containing colourless liquids.

One label states: 'Water, Boiling point 100°C, Density 1.0 g mL^{-1}'.

The other states: 'Hept-1-yne, Boiling point 100°C, Density 0.8 g mL^{-1}'.

Outline a method for identifying the contents of each bottle.

_____

_____

_____

_____

_____

3 The flash point of a hydrocarbon is the lowest temperature at which a volatile mixture of the volatile hydrocarbon and air can be ignited with a flame.

The ignition temperature of a hydrocarbon is the lowest temperature at which a mixture of a volatile hydrocarbon and air can spontaneously ignite.

The table shows the boiling points, flash points and ignition temperatures of the first eight members of the homologous series of straight chain alkanes.

Number of carbon atoms	Alkane	Boiling point (°C)	Flash point (°C)	Ignition temperature (°C)
1	Methane	−163	−135	595
2	Ethane	−88	−135	515
3	Propane	−42	−104	470
4	Butane	−0.5	−60	365
5	Pentane	36	−49	260
6	Hexane	69	−20	230
7	Heptane	98	−7	220
8	Octane	125	12	205

a Draw graphs for the number of carbon atoms versus the boiling point, flash point and ignition temperature.

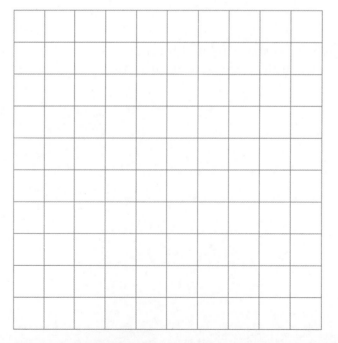

b Explain the trends in the graphs.

_____

_____

_____

_____

c Butane is used in gas lighters. A combustible mixture forms when there is a 1.8–8.4% mixture of gaseous butane and air. A 15 mL sample of gaseous butane is mixed with 200 mL of air in a container. The container is heated to 100°C.

    i Is the mixture a combustible mixture?

_____

_____

    ii Will the mixture spontaneously ignite in the absence of a flame at 100°C?

_____

_____

_____

_____

    iii Will the mixture ignite if a match is brought near the mixture?

_____

_____

WS 9.2 **Examining uses and disposal of organic substances**

LEARNING GOALS

Describe the processes for breaking up long chain hydrocarbons and identify the products

Assess the impact of a major oil spill

Identify safety symbols and their meanings

Investigate the disposal of organic substances

1 In the petrochemical industry, there is a greater demand for shorter chain hydrocarbons than for longer chain hydrocarbons.

   a State the names of the two processes that break longer chain hydrocarbons into shorter chain hydrocarbons.

   _____

   b Explain the difference between the two processes you identified in part a.

   _____

   _____

   _____

   _____

   _____

2 A student set up the equipment shown below in the laboratory to crack the liquid hydrocarbon $C_{12}H_{26}$ into a shorter chain hydrocarbon. She bubbled the gas formed, X, through a colourless liquid, Y.

Solid catalyst

Boiling tube

Liquid hydrocarbon in mineral wool

X

Y

   a Suggest an identity for liquid Y. Give a reason for your response.

   _____

   _____

   _____

   b The student predicted that X is pure ethene because it is a gas. Write an equation for the catalytic cracking of $C_{12}H_{26}$ to form ethene and another product.

   _____

   c Evaluate the student's prediction that X is pure ethene with reference to your answer for part b.

   _____

3 The diagram below shows the layers in which oil, gas and water are found underground.

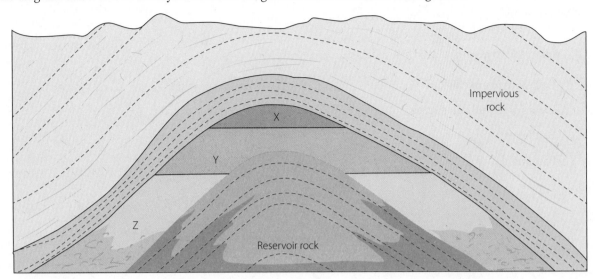

a Identify the layers where oil, gas and water are found as X, Y and Z. Justify your response.

_____

_____

_____

_____

b Assess the impact of mining and transportation of oil and gas on society and the environment with respect to the Deepwater Horizon disaster of 2010.

_____

_____

_____

_____

_____

_____

_____

_____

4 The Globally Harmonised System (GHS) is used to label chemicals so they can be handled and disposed of appropriately.

a Match the GHS symbols to their labels.

Corrosive chemicals

Environmentally damaging

Explosive

Flammable

Gases under pressure

Irritants, health hazard

Oxidiser

Toxic or poisonous

i

_____

ii

_____

iii

_____

iv

_____

v

_____

vi

_____

vii

_____

viii

_____

ix

_____

b The GHS labels for some organic compounds are given in the table. Complete the table to show how organic compounds should be handled and disposed of at the end of an experiment.

	Organic compound	GHS label(s)	Handling and disposal
i	Cyclohexane		
ii	Dichloromethane		
iii	Ethanol		
iv	Methanol		

INQUIRY QUESTION: WHAT ARE THE PRODUCTS OF REACTIONS OF HYDROCARBONS AND HOW DO THEY REACT?

WS **10.1** Investigating addition reactions

STUDENT BOOK
Pages 296–8

LEARNING GOALS

Investigate addition reactions of hydrogen, halogens, hydrogen halides and water to unsaturated hydrocarbons

Determine the reactants, given the products of addition reactions

1 All the reactions in the table below are examples of addition reactions. Complete the missing substances and, if appropriate, add the catalyst for the reactions listed. Write the IUPAC names for each organic compound in the table.

HINT

In some instances, there may be more than one product.

	Hydrocarbon		Reagent	Catalyst	Product(s) – names and formulae
a	$H_2C=CH_2$	+	HCl(g)	→	
b	$H_2C=CH-CH_3$	+		→	$CH_3-CH_2-CH_3$
c		+		→	$CH_3-CHCl-CHCl-CH_2-CH_3$ ... $H-C-C-C-C-CH_3$
d	$H-C\equiv C-H$	+	$2F_2$(g)	→	
e		+		→	CH_3-CH_2-O-H
f	$H-C\equiv C-H$	+	HCl(g)	$HgCl_2$ →	

2 A chemist performed two addition reactions but forgot to label the containers in which she collected the gases produced in the reactions. She knew one of the gases was chlorine and the other was steam with dilute sulfuric acid.

In order to identify the gases, she poured 20 mL of pent-1-ene into each of two 100 mL conical flasks labelled A and B. She then bubbled about 10 mL of one gas into one conical flask and about 10 mL of the other gas into the second flask.

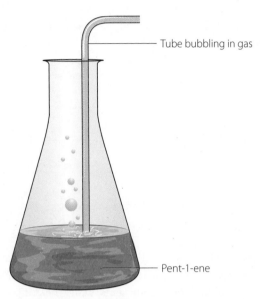

Tube bubbling in gas

Pent-1-ene

Then she poured each of the reaction mixtures into different separating funnels labelled A and B. She found that separating funnel A had two layers, which she labelled X and Y, while separating funnel B had only one layer, which she labelled Z.

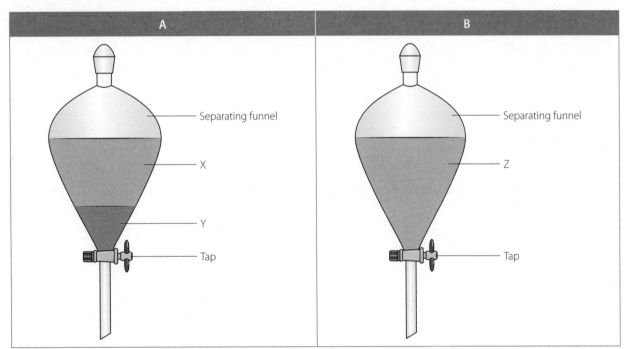

A	B
Separating funnel	Separating funnel
X	Z
Y	
Tap	Tap

a Write a balanced equation for each of the reactions of pent-1-ene with chlorine gas and with the steam with dilute sulfuric acid.

_____

_____

_____

_____

b Identify the contents of the layers labelled X, Y and Z.

X: _____

Y: _____

Z: _____

c State which gas was bubbled through each of the conical flasks A and B respectively. Justify your response with evidence from the experiment.

_____

_____

_____

_____

_____

_____

_____

3 Markovnikov was a Russian chemist in the nineteenth century. He predicted that when asymmetrical alkenes undergo addition reactions with a substance HX, where X can be a halide or hydroxyl group, the H is added to the carbon atom of the original double that had more hydrogens, while the X attaches to the carbon that had fewer hydrogens. There are therefore usually two products when asymmetrical alkenes react, with the major product being predicted by Markovnikov's rule.

Draw structural formulae and name the major and minor products in the following reactions.

	Reactants	Major product	Minor product
a	Propene + HCl(g)		
b	But-1-ene + H_2O (in the presence of dilute H_2SO_4)		

4 Identify the reactants that can be used to make only hexan-3-ol with no minor product.

_____

_____

Extension

5 Hydrogenation of two isomers with the formula C_4H_8, in the presence of a platinum catalyst, produces just one product. Draw and name the two isomers and the product formed. Explain why only one product is formed.

_____

_____

1 A substitution reaction of a hydrocarbon is shown below.

H   C   Cl

a Explain why the reaction shown above is classified as a substitution reaction. In your response include a balanced chemical equation.

_____

_____

_____

_____

_____

b Explain the role of the UV light shown on the arrow.

_____

c Explain why the hydrocarbon shown is referred to as 'saturated'.

_____

_____

_____

_____

2 Carbon tetrachloride, or tetrachloromethane, CCl_4, can be produced by substitution reactions of methane, CH_4. It is a colourless liquid that can be used as a solvent.

a Write the four steps for the substitution of each hydrogen in methane by chlorine.

_____

_____

_____

_____

b Write the net equation for the formation of CCl_4 from methane.

_____

3 A 3.007 g sample of ethane was reacted with excess bromine to produce 18.78 g of a haloalkane product and some hydrogen bromide gas.

a Identify the molecular formula of the product.

_____

_____

_____

_____

_____

_____

b Write a balanced equation for the substitution reaction.

_____

4 Write net equations using structural formulae for the reactions shown below that occur in the presence of UV light. Name all the products formed.

a Propane + bromine (in excess)

b Butane + chlorine (in a 1 : 1 mole ratio)

WS 10.3 Distinguishing between a saturated and an unsaturated hydrocarbon

Examine the reactions of a saturated and an unsaturated hydrocarbon with bromine water

Identify unknown samples as being saturated or unsaturated hydrocarbons based on experimental results and properties

1 The labels had fallen off two bottles containing colourless liquids. One label stated 'cyclohexane' and the other stated 'cyclohexene'. Students were asked to identify the contents of each bottle.

a A student added bromine water to samples of both, repeating the experiment three times but the bromine water did not change colour. Suggest a possible systematic error the student was making.

_____

b Outline a systematic method the student could have used to identify contents of each bottle.

_____
_____
_____
_____
_____
_____
_____

c Explain why two immiscible layers form when bromine water is added to the hydrocarbon and why it is important to shake the test tube.

_____
_____
_____
_____
_____

d Write an equation using structural formulae to show the hydrocarbon that reacted with bromine water.

e Give reasons why it is preferable to use cyclohexane and cyclohexene rather than ethane and ethene.

_____

_____

_____

_____

_____

f Explain how the waste should have been disposed of at the end of the experiment.

_____

_____

2 Two colourless hydrocarbon gases were reacted with bromine water in the absence of UV light. Only one of the gases reacted. The gas that reacted was bubbled through water and collected by the downward displacement of water in a 1 L measuring cylinder, as shown in experiment 1. The volume of gas collected at 25°C was 558 mL and the gas had a mass of 631 mg.

A fresh sample of the gas that reacted was then bubbled through dilute sulfuric acid, but this time no gas was collected, as shown in experiment 2.

a Describe the observation that led the experimenter to believe one of the gases reacted with bromine water.

_____

b How would the gas that did not react be described in terms of bonding?

_____

c Identify the gas that reacted with the bromine water.

_____

_____

_____

_____

_____

_____

d Explain why there was no gas collected in experiment 2. Write an equation using structural formulae in your response.

_____

_____

INQUIRY QUESTION: HOW CAN ALCOHOLS BE PRODUCED AND WHAT ARE THEIR PROPERTIES?

WS **11.1** Investigating alcohols

STUDENT BOOK
Pages 265, 316–325

LEARNING GOALS

Identify primary, secondary and tertiary alcohols

Name primary, secondary and tertiary alcohols

Identify intermolecular and intramolecular forces between and within alcohols and the effect of intermolecular forces on physical properties

1 Complete the table below.

	Structure	IUPAC name	Primary, secondary or tertiary alcohol
a		Butan-2-ol	
b	H—C—O—H (with H above and H below the C)		
c		2-Methylpropan-2-ol	
d	(cyclohexane ring)—OH		
e		2-Methylpentan-3-ol	

2 As the hydrocarbon chain length of the primary alcohols increases, from methanol to octan-1-ol, explain the trend in their solubility in a solvent such as hexane.

_____

_____

_____

_____

_____

3 The boiling points of the first six members of the homologous series of chloroalkanes and primary alcohols are given in the table.

Number of carbons	Chloroalkane	Boiling point (°C)	Alcohol	Boiling point (°C)
1	Chloromethane	−24	Methanol	65
2	Chloroethane	21	Ethanol	78
3	1-Chloropropane	46	Propan-1-ol	97
4	1-Chlorobutane	78	Butan-1-ol	117
5	1-Chloropentane	108	Pentan-1-ol	138
6	1-Chlorohexane	135	Hexan-1-ol	158

a Plot a graph of the boiling points versus the number of carbon atoms for each of the series listed above.

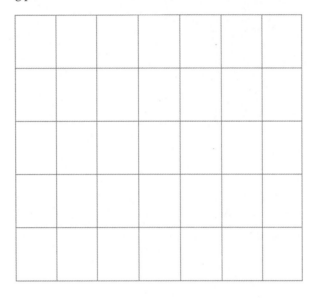

b Explain the trend of the boiling points shown in the graph for chloroalkanes and primary alcohols.

_____

_____

_____

_____

_____

_____

_____

_____

_____

_____

_____

_____

_____

_____

4 Pentan-1-ol, pentan-2-ol and 2-methylbutan-2-ol are all isomers with the molecular formula $C_5H_{11}OH$; that is, they all have the same number of atoms and the same molar mass. Explain whether they will all have the same boiling point. In your response, refer to the intermolecular and intramolecular forces in these compounds.

_____

_____

_____

_____

_____

_____

_____

_____

_____

_____

_____

_____

_____

_____

_____

5 The structures of two important hormones, oestradiol and testosterone, found in females and males respectively, are shown below.

Oestradiol	Testosterone
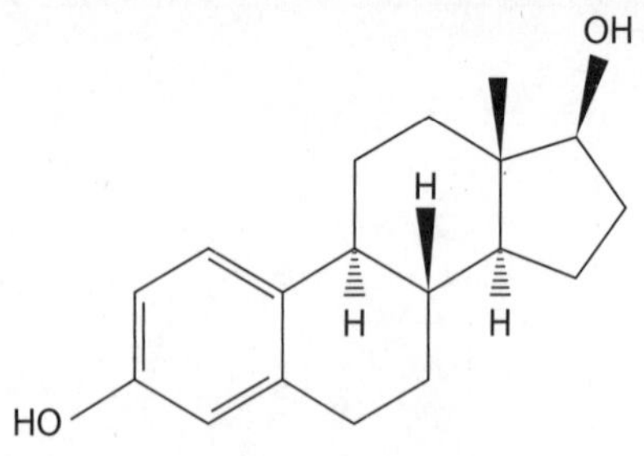	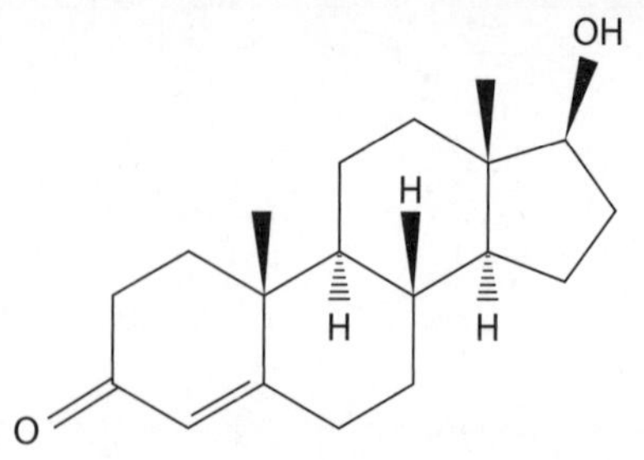
MM 272.4 g mol^{-1}	*MM* 288.4 g mol^{-1}

a Identify the functional groups present in each hormone.

_____

b Identify whether the alcohol groups are primary, secondary or tertiary in each compound.

_____

c The melting point of oestradiol is about 170°C, while that of testosterone is about 155°C. Explain the difference in the melting point with reference to intermolecular forces.

_____

_____

_____

_____

_____

_____

_____

_____

_____

_____

_____

_____

d Explain why the hormones are more soluble in ethanol than in water. Include a labelled diagram of ethanol in your response.

_____

_____

_____

_____

_____

Investigating enthalpy of combustion of alcohols

STUDENT BOOK
Pages 316–18

Calculate the enthalpy of combustion of alcohols

Analyse the validity, reliability and accuracy of data collected from an investigation

Apply evidence to identify an unknown

1 Two students, Ali and Sam, set up the following equipment to measure the enthalpy of combustion of propan-1-ol. The theoretical value for the molar enthalpy of combustion of propan-1-ol is 2021 kJ mol^{-1}.

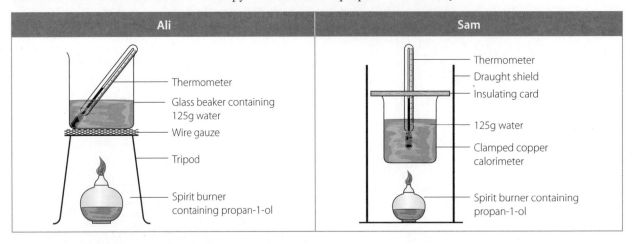

The data collected by Ali and Sam are given in the table.

	Ali	Sam
Initial mass (spirit burner + propan-1-ol) (g)	265.243	268.357
Final mass (spirit burner + propan-1-ol) (g)	264.884	268.148
Initial temp of water (°C)	22	22
Final temp of water (°C)	34	34
Mass (water) (g)	125	125
Observation	Bottom of the beaker covered in soot	Bottom of copper can had minimal soot

a Write an equation for the complete combustion of propan-1-ol.

_____

b Calculate the molar enthalpy of combustion obtained by:

i Ali

_____

_____

_____

_____

_____

_____

ii Sam

_____

_____

_____

_____

_____

c **i** Calculate the percentage error in Ali and Sam's experimental results.

_____

_____

_____

ii Account for the difference in percentage error of Ali and Sam's results with reference to the theoretical molar enthalpy of combustion of propan-1-ol. In your response refer to the validity, reliability and accuracy of each experiment.

_____

_____

_____

_____

_____

_____

_____

_____

_____

_____

_____

_____

2 A student was given three tins labelled X, Y and Z as shown below on the left, that were used to determine the molar enthalpy of combustion of three unknown alkanols. Approximately the same number of moles of each alcohol was used. The base of tin X had a moderate amount of soot, tin Y had very little soot, while the base of tin Z was covered in soot. The experimental set-up is also shown (on the right).

The student was told that the unknown alcohols were methanol, ethanol and pentan-1-ol. Match the tins labelled X, Y and Z to the alkanols, justifying your response with appropriate equations.

_____

_____

_____

_____

_____

_____

_____

_____

_____

_____

_____

LEARNING GOALS

Investigate reactions of alcohols that include combustion, dehydration, substitution with hydrogen halides and oxidation

Write equations for reactions of alcohols

Predict products of reactions of alcohols

Determine the identity of an alcohol from combustion data

Design an experimental method to identify unknown alcohols

1 Write balanced equations for the following reactions and name the organic product where appropriate.

   a Complete combustion of ethanol

   _____

   b Incomplete combustion of butan-1-ol to produce soot, carbon monoxide and water

   _____

   c Dehydration of propan-2-ol

   _____

   d Substitution of methanol with hydrogen chloride gas

   _____

   e Oxidation of hexan-2-ol using an oxidising agent represented by

   $$\xrightarrow{Cr_2O_7{}^{2-}/H^+}$$

   _____

   f Oxidation of 2-methylpropan-2-ol using an oxidising agent represented by $[O_x]$

   _____

2 An alcohol was first dehydrated and then reacted with bromine water. There were two products formed: 1,2-dibromobutane and 2,3-dibromobutane. Name the alcohol used and explain, using balanced equations to support your answer, how the identity of the alcohol is determined.

   _____

   _____

   _____

   _____

   _____

3 Complete combustion of 0.2203 g of a primary alcohol produced 0.5501 g of carbon dioxide and 0.2702 g of water. Identify the primary alcohol.

_____

_____

_____

_____

_____

_____

_____

_____

_____

_____

_____

_____

_____

_____

4 Bromoethane can be produced by adding the same non-organic reagent to two different organic molecules in an addition reaction and a substitution reaction. Write equations using structural formulae to show each reaction.

5 The labels have fallen off three bottles containing colourless liquids.

The labels read: ethanol, 2-methylpropan-2-ol and octan-1-ol.

a Outline a method for identifying the contents of each bottle.

b Explain your expected observations in part a.

LEARNING GOALS

Evaluate experiments designed to produce ethanol by fermentation

Investigate the production of alcohols using substitution reactions of haloalkanes and addition reactions of alkenes

Apply reactions of alcohols to identify products

1 Using the same reaction mixture, two students set up different experiments for the fermentation of glucose, labelled X and Y as shown below.

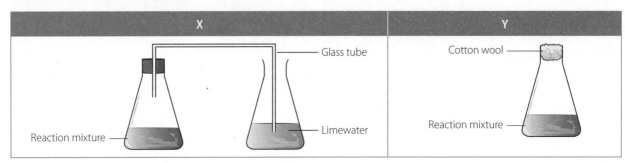

a Identify the components in the reaction mixture.

_____

b Write an equation for the fermentation of glucose.

_____

c State the conditions required for the fermentation of glucose.

_____

_____

d The students recorded the mass of their reaction flask over five days. The data collected are shown below.

Day	Mass of flask X (g)	Mass of flask Y (g)
1	250.12	247.15
2	245.63	247.14
3	243.16	246.11
4	243.16	246.10
5	243.16	245.95

i Explain the data observed for each flask with reference to the fermentation process. Suggest an improvement in either experiment to support your explanation.

_____

_____

_____

_____

_____

_____

_____

_____

_____

_____

ii Explain why limewater was used in procedure X.

_____

_____

iii Calculate the percentage of glucose that was fermented based on 97.57 g glucose being placed in the reaction mixture.

_____

_____

_____

_____

_____

_____

_____

_____

_____

_____

e Repeating procedure X several times showed that no more than 15% glucose could be fermented. Explain why this was observed.

_____

_____

f Describe a procedure for separating the ethanol from the fermentation mixture.

_____

_____

2 Ethanol can be produced by fermentation, as seen in question 1. It can also be formed by addition and substitution reactions of an alkene and haloalkane respectively. Write equations to show these reactions for the production of ethanol.

Addition reaction: _____

_____

Substitution reaction: _____

_____

3 An alcohol with molecular formula $C_4H_{10}O$ could not be oxidised but it was produced from a substitution reaction of a haloalkane. Identify the alcohol, write the equation for its production and name the organic reactant. Justify your response.

_____

_____

LEARNING GOALS

Write half-equations and net redox equations for the oxidation of alcohols

Identify the products formed from the oxidation of alcohols

Identify experimental equipment and analyse experimental set-up

Evaluate experimental results

HINT

When writing half-equations for oxidation and reduction, the following steps may be used.

1 Balance for atoms other than oxygen and hydrogen
2 Balance for oxygen by adding H_2O to the appropriate side
3 Balance for hydrogen by adding H^+ to the appropriate side
4 Balance for charge by adding e^- to the appropriate side

---

1 Ethanol can be oxidised to ethanal by acidified potassium dichromate using the set-up shown below.

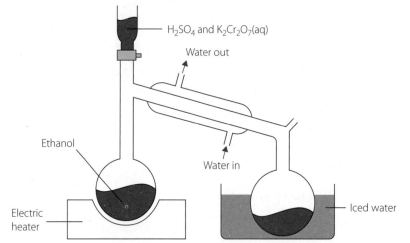

a Explain how it is possible to collect the ethanal using the above set-up.

_____

_____

_____

b Explain why the ethanol cannot be oxidised to ethanoic acid using the above set-up.

_____

_____

c Explain why an electric heater is used to heat the reaction mixture rather than a Bunsen burner.

_____

_____

d Write the oxidation and reduction half-equations and the net equation to show the oxidation of ethanol to ethanal using acidified potassium dichromate.

Oxidation half-equation: _____

Reduction half-equation: _____

Net equation: _____

e Identify and explain a colour change that would be observed in the reaction flask.

_____

f Identify the following in the above reaction.

Oxidant: _____

Reductant: _____

2 Propan-1-ol can be oxidised to propanoic acid using the set-up shown below.

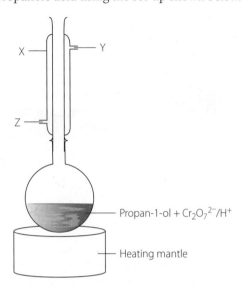

a Identify the names for labels X, Y and Z.

X: _____

Y: _____

Z: _____

b Justify how propanoic acid can be produced from propan-1-ol using the set-up shown above.

_____

_____

_____

_____

c Write the half-equations for oxidation and reduction and the net equation for the reaction.

Oxidation half-equation: _____

Reduction half-equation: _____

Net equation: _____

d Suggest one more item that needs to be added to the reaction flask and the reason for adding it.

_____

3 The label had fallen off a bottle containing a colourless liquid. The label was partially destroyed, and it was not possible to determine whether it read 'Butan-1-ol' or 'Butan-2-ol'. A student decided to oxidise the unknown alcohol using acidified potassium permanganate in the set-up shown.

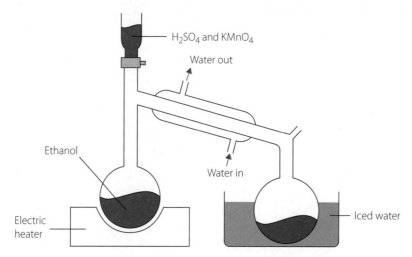

a The student observed a change in colour of the KMnO$_4$ as the reaction proceeded. State the colour change observed.

_____

_____

b The student concluded that butanal had been produced in the above reaction and that the label should have read 'Butan-1-ol'. She determined the boiling point of the product in order to confirm her conclusion. The literature search revealed that the boiling point of butanal is 75°C. However, repeated tests by the student yielded a boiling point of 80°C. The student concluded that there must have been a systematic error in her results. Evaluate the student's conclusion in terms of validity. Suggest a further test the student can perform to support her conclusion.

_____

_____

_____

_____

_____

_____

Identify the two main biofuels used in Australia

Identify the structure of biodiesel and the reaction for the production of biodiesel

Compare fossil fuels to biofuels in terms of CO_2 emission and enthalpy of combustion

1 a Define the term 'fuel'.

_____

b Fossil fuels currently provide most of Australia's energy. Give two examples of fossil fuels and explain current concerns with their use.

_____

_____

_____

_____

_____

c Define the term 'biofuel' and give examples of the two main biofuels used in Australia.

_____

_____

_____

_____

2 Diesel is a type of fossil fuel. What are the differences between the chemical structures of diesel and biodiesel?

_____

_____

_____

3 a Identify the name of the reaction that converts vegetable oils into biodiesel.

_____

b Identify the parts of the molecule shown, X, Y and Z, using the following terms: glycerol, saturated fatty acids, triglyceride.

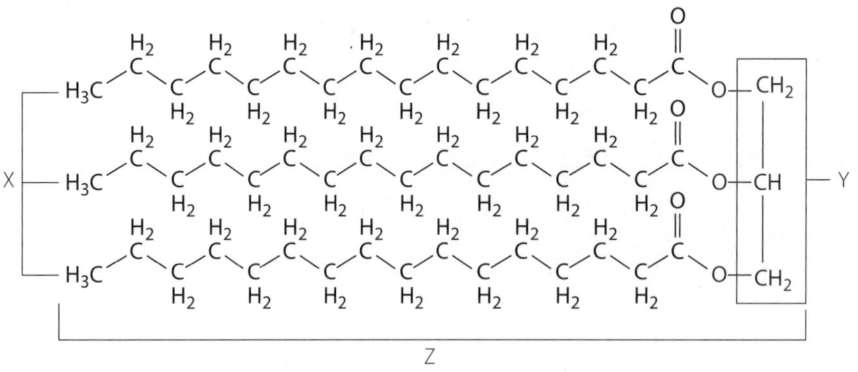

X: _____

Y: _____

Z: _____

c Complete the missing parts of the equation to show the formation of a biodiesel.

Reactant 1	Reactant 2	Conditions	Product 1	Product 2
		$\longrightarrow$		

d Enzymes such as lipase may be used to catalyse the reaction in part c, instead of NaOH or KOH. Discuss the use of these reagents to catalyse the reaction.

_____

_____

_____

_____

_____

_____

4 The fuel tank capacity of a car is 60 L.
Consider the data provided.

Physical property	Bioethanol	Octane (petrol)
Density (g mL^{-1})	0.78	0.69
Energy (kJ g^{-1})	29.6	48

a Compare the volume of carbon dioxide produced at 25°C and 100 kPa by the complete combustion of two 60 L fuel tanks, one of which contains bioethanol and the other contains octane.

_____

_____

_____

_____

_____

_____

_____

_____

_____

b Calculate the volume of bioethanol that would be required to produce the same amount of energy as released by 1 g of octane.

_____

_____

_____

_____

12 Reactions of organic acids and bases

WS 12.1 Investigating properties of functional groups

STUDENT BOOK
Pages 263–84, 342–51

LEARNING GOALS

Identify products of reaction sequences

Explain properties of alcohols, aldehydes, ketones, amines, amides and carboxylic acids

Identify functional groups in a compound

Perform mole calculations based on data

Construct a graph and analyse trends

1 Complete the missing items in the boxes labelled 1–7 and reagent X in the flow chart below, using the terms from the list, and draw the structural formulae where appropriate. The terms may be used once, more than once or not at all.

$H^+/Cr_2O_7^{2-}$	butan-1-one	butanamine
1-chlorobutane	butan-2-ol	butanoic acid
but-1-ene	butan-2-one	concentrated H_2SO_4
but-2-ene	butanal	dilute H_2SO_4
butan-1-ol	butanamide	

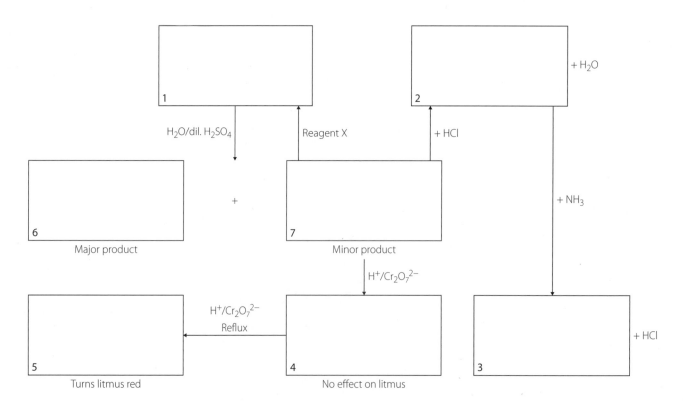

2 a Draw structural formulae of pentanamine, *N*-methylpropanamine and *N,N*-dimethylpropanamine in the space below.

b Rank the compounds in part a in increasing order of boiling point, explaining your response.

_____

_____

_____

_____

_____

_____

_____

_____

3 Lactam is the term used to refer to cyclic amides. A β-lactam is a four-membered ring, as shown in the diagram on the left. Penicillins are a group of compounds referred to as β-lactam antibiotics. The structure of amoxicillin, a common antibiotic, is shown in the diagram on the right.

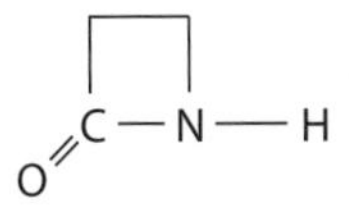

β-lactam 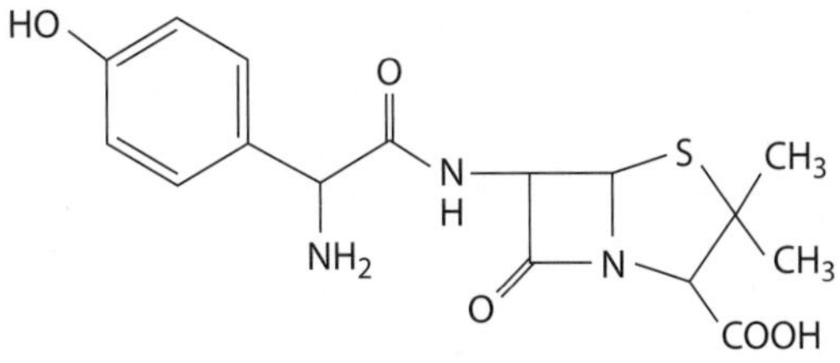   Amoxicillin

a Identify all the functional groups in amoxicillin by circling and naming the groups in the figure above.

b Explain whether amoxicillin is soluble in water.

_____

_____

_____

_____

_____

4 During starvation, ketones are formed in the body from the breakdown of fatty acids. A fatty acid is a carboxylic acid that has a long hydrocarbon chain. The graph shows the variation in levels of propanone, CH_3COCH_3, and stearic acid (or octadecanoic acid), $CH_3(CH_2)_{16}COOH$, during starvation. The unit mM refers to millimoles per litre.

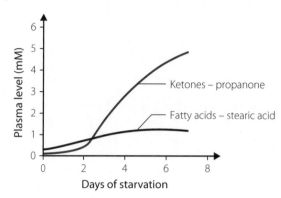

a Calculate the mass of propanone and stearic acid, in milligrams, that would be present in a 10 mL plasma sample taken on the fifth day of starvation.

_____

_____

_____

_____

_____

_____

_____

_____

b Explain why the solubility of propanone in water is higher than that of stearic acid in water.

_____

_____

_____

_____

_____

5 The boiling points of some aldehydes, ketones, amides and carboxylic acids are listed below.

Number of carbons	Aldehyde B.P. (°C)	Ketone (-2-one) B.P. (°C)	Amide B.P. (°C)	Carboxylic acid B.P. (°C)
1	−21		193	101
2	21		222	118
3	46	56	213	141
4	75	80	216	164
5	103	102	225	186

a Explain why the first member of the ketone series has three carbon atoms.

_____

b Plot a graph of the boiling points.

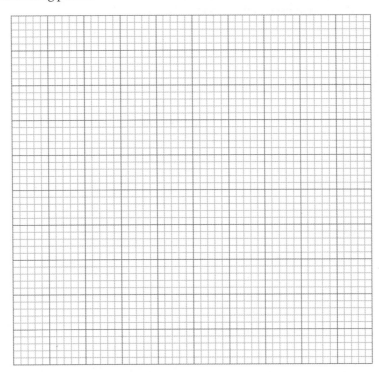

c Explain the trends in the boiling points shown in the graph.

_____

_____

_____

_____

_____

_____

_____

_____

_____

_____

LEARNING GOALS

Examine the equipment and materials used in the production of esters

Name esters using IUPAC nomenclature

Draw organic structural formula

Describe a method for identifying unknown samples

1 a Write a balanced equation using structural formulae to show the formation of butyl ethanoate. Write the names of the organic reactants in your response.

b The diagram below shows the preparation of butyl ethanoate.

i Match the terms in the list below with the labels 1 to 7.

Boiling chips (porcelain)   Reflux condenser

Cold water in from tap   Round-bottom flask

Heating mantle   Water out to sink

Reaction mixture

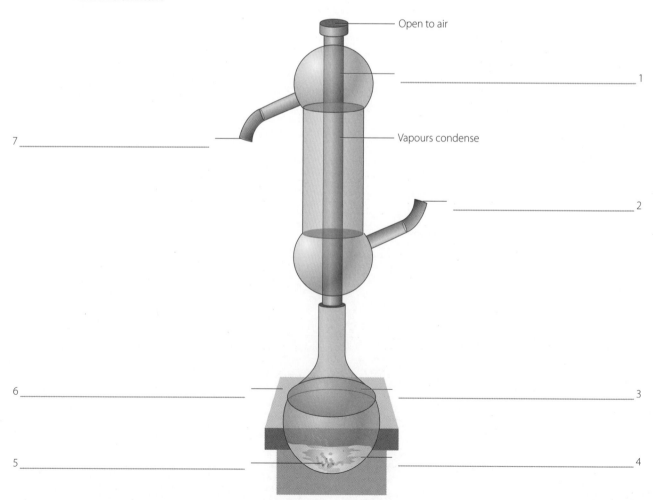

Open to air

1

Vapours condense

7

2

6 _____ 3

5 _____ 4

ii State the name of the process used to make the ester butyl ethanoate and justify why the set-up shown above is used.

_____

_____

_____

_____

_____

_____

_____

_____

_____

c The above reaction was allowed to take place for one hour. Identify the compound(s) present in the reaction mixture at the end of this time.

_____

The butyl ethanoate ester was purified using three steps.

d The first step to separate the ester involved transferring the reaction mixture to the equipment shown and then adding water.

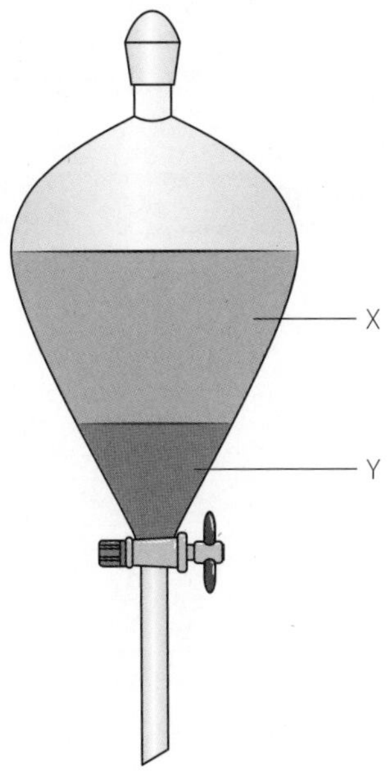

i Name the equipment shown.

_____

ii Explain why two immiscible layers form and identify the contents of the layers labelled X and Y.

_____

_____

_____

e In the second step, the water layer was removed from the separating funnel and aqueous sodium carbonate was added. Two layers formed again, and this time bubbling was observed. Explain the need to add the aqueous sodium carbonate solution. Write a balanced equation in your response.

_____

_____

f Identify the name of the procedure used in the final step to purify the ester.

_____

2 Three 100 mL beakers were half filled with some chemicals but the beakers were not labelled. The chemicals used were ethanol, ethyl methanoate and methanoic acid. Suggest a method by which the contents of the beakers can be identified.

_____

_____

_____

_____

_____

_____

_____

_____

3 Complete the missing sections of the table, with names and structural formulae, to show the formation of esters. Water is formed in all the reactions.

	Alkanol	Carboxylic acid	Ester	Water
a			Methyl hexanoate	
b	Propan-1-ol		Propyl butanoate	
c		Pentanoic acid	Ethyl pentanoate	

LEARNING GOALS

Investigate the formation and action of soaps and detergents

Write balanced equations for the reaction used in the production of soap

1 Define the term 'saponification'.

_____

_____

2 Complete the equation to show how the triglyceride given below can be used to make soap. Identify the soap and the other organic product in the reaction.

$$
\begin{array}{l}
\text{H} \quad\quad \text{O} \\
| \quad\quad\quad || \\
\text{H} - \text{C} - \text{O} - \text{C} - (\text{CH}_2)_{16} - \text{CH}_3 \\
| \quad\quad\quad\quad\quad \text{O} \\
| \quad\quad\quad\quad\quad || \\
\text{H} - \text{C} - \text{O} - \text{C} - (\text{CH}_2)_{16} - \text{CH}_3 \quad + \quad\quad\quad\quad\quad\quad \longrightarrow \quad\quad\quad\quad\quad\quad\quad + \\
| \quad\quad\quad\quad\quad \text{O} \\
| \quad\quad\quad\quad\quad || \\
\text{H} - \text{C} - \text{O} - \text{C} - (\text{CH}_2)_{16} - \text{CH}_3 \\
| \\
\text{H}
\end{array}
$$

Triglyceride

3 Soap, with a general formula $RCOO^-Na^+$, is not effective in acidic solutions with a pH less than about 4.5 nor is it effective in hard water that contains calcium and/or magnesium ions. Explain, using relevant equations, why soap is ineffective in these solutions.

_____

_____

_____

_____

_____

_____

4 Explain the cleaning action of soap with reference to the diagram shown.

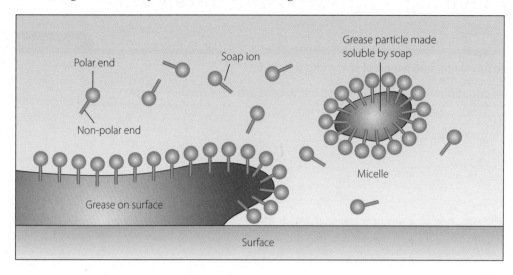

_____

_____

_____

_____

_____

_____

_____

_____

_____

_____

_____

_____

5 a Label the detergents shown as anionic, cationic or non-ionic.

Detergent	Anionic, cationic or non-ionic
i $$\left[CH_3(CH_2)_{15} - \overset{\overset{\displaystyle CH_3}{\displaystyle |}}{\underset{\underset{\displaystyle CH_3}{\displaystyle |}}{N}} - CH_3 \right]^+ Br^-$$	
ii $$CH_3(CH_2)_{10}CH_2 - \overset{O}{\underset{O}{\overset{\|}{\underset{\|}{S}}}} - ONa$$	
iii	

b For each type of detergent listed below, suggest a use based on a property.

i Anionic

_____

_____

ii Cationic

_____

_____

_____

_____

iii Non-ionic

_____

_____

6 Students were given two 250 mL beakers, each half filled with two liquids. They were told one of the beakers contained a soap solution, while the other beaker contained a cationic detergent. Outline a procedure for identifying the contents of each beaker. Justify your response.

_____

_____

_____

_____

_____

_____

_____

_____

_____

WS 13.1 Investigating addition polymers

STUDENT BOOK
Pages 368–84

LEARNING GOALS

Describe the structures and uses of polyethylene, polyvinyl chloride, polystyrene and polytetrafluoroethylene

Identify monomers and their corresponding polymers

Calculate the number of fluorine atoms in a polymer sample

Represent a chemical process as a flow chart

Relate properties of polymers to structure

Represent a polymer production reaction

1 State what feature a monomer needs to have in order to form an addition polymer.

_____

2 Complete the following table.

Monomer name		Monomer structure	Polymer repeat unit structure	Polymer name	Use
Common	Systematic				
Ethene					
Vinyl chloride					
	Ethenylbenzene				
	Tetrafluoroethylene				

3 A sample of polytetrafluoroethene (PTFE) has an average molar mass of $5.5511 \times 10^4 \, \text{g mol}^{-1}$. Calculate the number of fluorine atoms in an average molecule.

_____

_____

_____

_____

_____

4 The figures labelled X and Y show the arrangement of polymer chains in two types of polyethene.

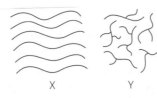

X   Y

a Identify the types of polyethene represented by X and Y and give a reason for your response.

_____

_____

_____

_____

_____

b Compare the flexibility and melting points of the structures labelled X and Y.

_____

_____

_____

_____

_____

c A sample of structure Y was analysed and found to contain a variety of molecular masses. Explain.

_____

_____

5 Complete the table to show the structures of a repeat unit for the polymers formed from the given monomers.

	Monomer	Polymer repeat unit
a	H−C(H)(H)−C(H)=C(H)(H)	
b	H(H)C=C(H)−C≡N	
c	$CH_2=CH-O-CO-CH_3$	

6 Draw a flow chart with equations to show the formation of polyvinylchloride (PVC) from crude oil, by referring to the paragraph.

Ethene is fractionally distilled from crude oil or obtained by catalytic cracking of longer chain hydrocarbons such as decane and separated from the mixture. Ethene is then reacted with chlorine gas using iron(III) chloride catalyst and the reaction is exothermic, $\Delta H = -178\,kJ\,mol^{-1}$. The product of this reaction is cracked by passing it through hot metal tubes at 650 K to produce hydrogen chloride and chloroethene. The hydrogen chloride is removed from the mixture. The chloroethene is polymerised using an organic initiator at a pressure of about 1300 kPa to ensure the monomer is kept in the liquid state. The PVC precipitates out as it forms because it is not soluble in the monomer.

Crude oil

↓

WS 13.2 Investigating condensation polymers

LEARNING GOALS

Compare the structures of polyamides and polyesters

Identify monomers and the corresponding condensation polymers

Distinguish between synthetic and naturally occurring polymers

Use data to calculate average mass of polymer formed

1 The structures of some compounds labelled A–F are given.

A	B	C	D	E	F

Select from the compounds provided above to show the formation of the following polymers from the monomers by writing equations using structural formulae, showing the appropriate linkage between monomer units.

a A polyamide

b A polyester

2 Complete the table to show the structure of the monomer(s) or polymer repeat unit.

	Monomer(s)	Polymer repeat unit
a		$\left[\begin{array}{c} \text{COOH} \quad\quad \text{COOH} \quad\quad \text{COOH} \\ \text{O-CH-CH}_2\text{-C-O-CH-CH}_2\text{-C-O-CH-CH}_2\text{-C} \\ \text{O} \quad\quad\quad \text{O} \quad\quad\quad \text{O} \end{array} \right]$
b		$\left[\begin{array}{c} \text{H} \quad \text{R} \quad\quad \text{H} \quad\quad \text{O} \\ \text{C} \quad\quad\quad \text{C} \\ \text{N} \quad\quad \text{C} \quad\quad \text{N} \quad\quad \text{C} \\ \text{H} \quad\quad \text{O} \quad\quad \text{R} \quad \text{H} \end{array} \right]$
c		$\left[\begin{array}{c} \text{O} \quad \text{O} \quad\quad \text{O} \quad \text{O} \quad\quad\quad \text{O} \\ \text{C-}\bigcirc\text{-C-NH-}\bigcirc\text{-C-}\bigcirc\text{-C-NH-}\bigcirc\text{-NH-C} \end{array} \right]$
d	$H_2N-(CH_2)_6-NH_2 + HO-\overset{O}{\overset{\|}{C}}-(CH_2)_4-\overset{O}{\overset{\|}{C}}-OH$	

3 Proteins are naturally occurring polymers that are formed from condensation polymerisation of amino acids. Amino acids contain both a carboxylic acid group and an amine group. The link is referred to as a peptide link. A polypeptide forms when many amino acids undergo condensation polymerisation.

The simplest amino acid is glycine. Its structure is given below.

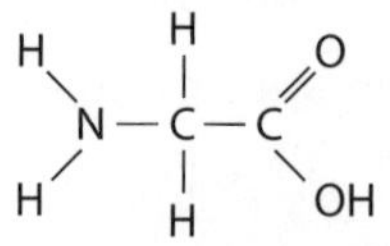

Alanine is another amino acid and its structure is given below.

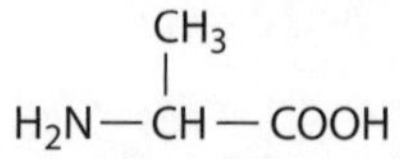

a Draw the structure of a polypeptide formed when n moles of alanine reacts with n moles of glycine and label the peptide link.

b Calculate the average mass of the polypeptide formed when 600 moles each of glycine and alanine polymerise.

_____

_____

_____

_____

_____

_____

_____

_____

_____

4 Nylon 6,6 is formed when hexane-1,6-diamine reacts with hexanedioyl dichloride. The structures of the two monomers are shown.

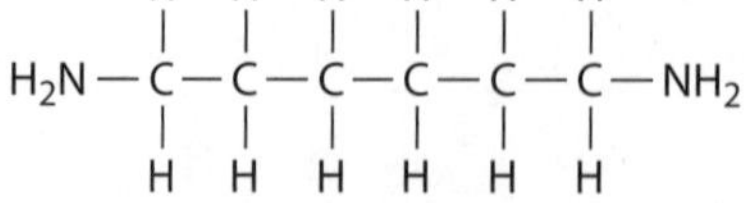

Hexane-1, 6-diamine 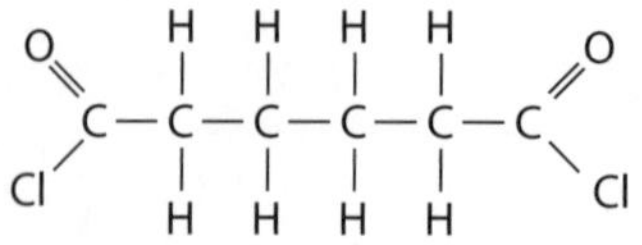   Hexanedioyl dichloride

a Write an equation to show the formation of part of the nylon 6,6 polymer when n moles of each monomer react and show the amide link.

b Determine the mass of the polymer formed when 18 229 g of hydrogen chloride forms.

_____

_____

_____

_____

_____

_____

_____

_____

_____

5 Condensation polymers can be 'synthetic' or 'naturally occurring'. Distinguish between these terms and provide examples to illustrate their use and biodegradability.

_____

_____

_____

_____

_____

_____

_____

_____

Module seven: Checking understanding

1 What is the IUPAC nomenclature for the compound shown?

$$\begin{array}{cccccc}
H & H & H & H & Cl & H \\
| & | & | & | & | & | \\
H-C- & C- & C- & C- & C- & C-H \\
| & | & | & | & | & | \\
H & H & Br & H & H & H
\end{array}$$

A 3-Bromo-5-chlorohexane

B 5-Chloro-3-bromohexane

C 4-Bromo-2-chlorohexane

D 2-Chloro-4-bromohexane

2 What term best describes the structures shown?

$$\begin{array}{ccccc}
H & H & H & H & H \\
| & | & | & | & | \\
H-C- & C- & C- & C- & C-H \\
| & | & | & | & | \\
H & H & H & H & H
\end{array}$$

$$\begin{array}{cccc}
 & H & & \\
 & | & & \\
 & H-C-H & & \\
H & | & H & H \\
| & | & | & | \\
H-C- & C- & C- & C-H \\
| & | & | & | \\
H & H & H & H
\end{array}$$

A They are isomers.

B They are chain isomers.

C They are position isomers.

D They are functional group isomers.

3 What is the IUPAC nomenclature for the compound shown by the ball and stick model?

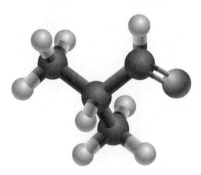

A Butanal

B Butan-1-one

C Methylpropanone

D Methylpropanal

4 What is the IUPAC nomenclature for the compound shown?

$$
\begin{array}{c}
\quad\; H \quad H \\
\quad\; | \qquad | \\
H-C-C-C \overset{\displaystyle O}{\diagup} \\
\quad\; | \qquad | \qquad\quad H \\
\quad\; H \quad H \quad\;\; N-C-H \\
\qquad\qquad H-C-H\;\; H \\
\qquad\qquad H-C-H \\
\qquad\qquad\quad\;\; H
\end{array}
$$

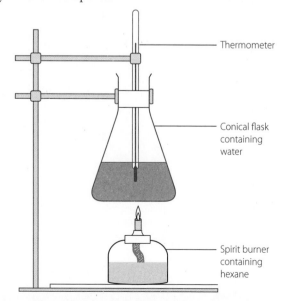

A *N,N*-Propyl propanamide

B *N,N*-Diethyl butanamide

C *N*-Ethylmethylpropanamide

D *N*-Ethyl-*N*-methylpropanamide

5 Which alternative correctly identifies the reactant and product of an *addition* reaction?

A

B

C

D

6 A Chemistry teacher used the equipment set-up shown to measure the enthalpy of combustion of hexane, heptane and octane so they could be compared.

- Thermometer

- Conical flask containing water

- Spirit burner containing hexane

Statements

  I The independent variable is water.

  II The independent variable is fuel.

  III The measured enthalpy of combustion will be greater than theoretical because the glass conical flask is a good conductor of heat.

  IV The measured enthalpy of combustion will be less than theoretical because the glass conical flask is a poor conductor of heat.

  V Soot is most likely to form at the bottom of the conical flask.

  VI Soot is unlikely to form at the bottom of the conical flask.

Which of the above statement(s) is/are correct?

A I, IV & V

B II, III & VI

C II, IV & V

D IV only

7 Which reagent can be used to convert ethanol to ethanoic acid?

  A Dilute sulfuric acid

  B Concentrated sulfuric acid

  C Potassium dichromate

  D Acidified potassium dichromate

8 Which option correctly identifies the name of the substance shown and the reactants used to make it?

	Name	Reactants	
A	Hexyl methanoate	Hexanol	Methanoic acid
B	Hexyl methanoate	Methanol	Hexanoic acid
C	Methyl hexanoate	Methanol	Hexanoic acid
D	Methyl hexanoate	Hexanol	Methanoic acid

9 What is the monomer unit for the polymer unit shown?

10 A student set up the equipment shown below to calculate the molar enthalpy of combustion of ethanol. The data given in the table was collected by the student. The student did not have access to an electronic balance to record the mass of the spirit burner, so they recorded the volume of ethanol used. The density of ethanol is $0.79\,g\,mL^{-1}$ and that of water is $1\,g\,mL^{-1}$.

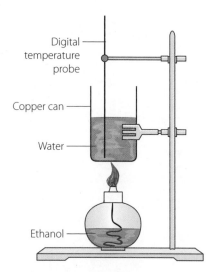

Measurement	Data
V(ethanol) initial (mL)	341.26
V(ethanol) final (mL)	335.65
V(water) (mL)	255.00
Temp (water) initial (°C)	22.64
Temp (water) final (°C)	43.87

a   Calculate the molar enthalpy of combustion of ethanol based on the student's data.

b   The theoretical molar enthalpy of combustion of ethanol is $1367\,kJ\,mol^{-1}$. Calculate the percentage error in the student's result. Account for the difference and suggest two improvements.

11 Propan-2-ol was oxidised to propanone using acidified potassium permanganate solution.

a Write the oxidation half-equation.

_____

b Write the reduction half-equation.

_____

c Write the NET equation.

_____

d State one observation for the reaction stated in part c.

_____

e Identify the oxidant in part c.

_____

f Identify the reductant in part c.

_____

12 A student was given two colourless liquids in beakers and asked to identify them as methanamide or methanamine. Outline how they would identify the liquids.

_____

_____

_____

_____

_____

13 Kevlar is used to make bulletproof vests. Its structure is shown below.

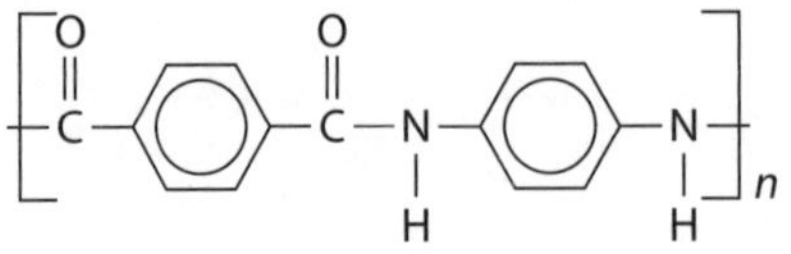

a Draw the structure(s) for the monomer(s) if hydrogen chloride is produced in the polymerisation reaction.

_____

_____

_____

b Plasticisers are often added to polymers. Explain the role of plasticisers.

_____

_____

_____

14 A compound labelled X is shown below.

Compound X

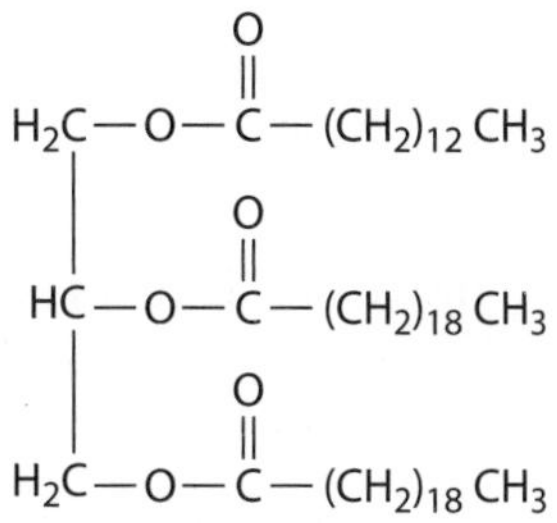

$$H_2C-O-\overset{\overset{\displaystyle O}{\|}}{C}-(CH_2)_{12}\,CH_3$$

$$HC-O-\overset{\overset{\displaystyle O}{\|}}{C}-(CH_2)_{18}\,CH_3$$

$$H_2C-O-\overset{\overset{\displaystyle O}{\|}}{C}-(CH_2)_{18}\,CH_3$$

a State the general name of the compound labelled X.

_____

b Complete the equation below to show the acid hydrolysis of compound X by refluxing.

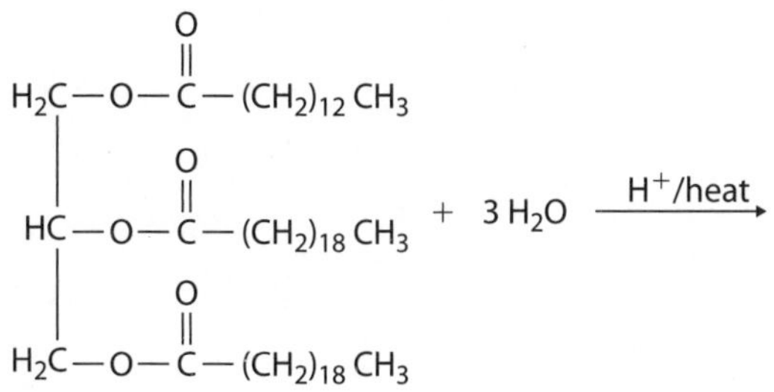

$$H_2C-O-\overset{\overset{\displaystyle O}{\|}}{C}-(CH_2)_{12}\,CH_3$$

$$HC-O-\overset{\overset{\displaystyle O}{\|}}{C}-(CH_2)_{18}\,CH_3 \quad + \quad 3\,H_2O \quad \xrightarrow{\;H^+/heat\;}$$

$$H_2C-O-\overset{\overset{\displaystyle O}{\|}}{C}-(CH_2)_{18}\,CH_3$$

15 An organic compound, X, containing carbon, hydrogen and nitrogen, turned litmus blue. When compound X was reacted with ethanoic acid and tested again with litmus, the product, Y, had no effect on litmus.

On combustion, 2.360 g of organic compound X produced 5.28 g of carbon dioxide and 3.24 g of water along with an oxide of nitrogen.

a Calculate the empirical formula of the organic compound X.

_____

_____

_____

_____

_____

_____

b A 2.360 g sample of the organic compound, X, was found to occupy a volume of 1.1888 L at 85°C and 100 kPa. Calculate the molar mass of X.

_____

_____

_____

_____

_____

c **i** Determine the structural formula of compound X and give its name.

ii Describe how you can accurately determine the identity of compound X.

_____

_____

_____

_____

d Suggest a possible structure and name for the product, Y.

Reviewing prior knowledge

1 What is solubility a measure of?

_____

_____

2 What is the difference between cations and anions?

_____

3 What happens in a precipitation reaction?

_____

4 What is the difference between a complete ionic equation and a net ionic equation?

_____

_____

5 Explain, using Le Chatelier's principle, whether the equilibrium constant for each of the following reactions increases or decreases as temperature rises.

a $HOCl(aq) + H_2O(l) \rightleftharpoons H_3O^+(aq) + OCl^-(aq)$

_____

b $PCl_5(g) + 93\ kJ \rightleftharpoons PCl_3(g) + Cl_2(g)$

_____

_____

6 Calcium sulfate, ($CaSO_4$), is a sparingly soluble salt.

a Write the equation for the dissociation of $CaSO_4$ in water.

_____

b Write the K_{sp} expression for this substance.

_____

c Explain the effect on the solubility of $CaSO_4$ of adding each of the following compounds to equilibrium solutions of $CaSO_4$.

  i Solid Na_2SO_4

  ii Solid KNO_3

  iii Solid $Ca(NO_3)_2$

_____

_____

_____

7 A solution contains $0.010 \, mol \, L^{-1} \, Cl^-$ ions and $0.0010 \, mol \, L^{-1} \, CrO_4^{2-}$ ions. The K_{sp} of AgCl is 1.56×10^{-10} and the K_{sp} of Ag_2CrO_4 is 9.0×10^{-12}.

a At what concentration of Ag^+ will each of the precipitates start forming?

b Which of these compounds will start precipitating first when Ag^+ is added to the solution?

c What will be the concentration of Cl^- ions when Ag_2CrO_4 begins to precipitate?

8 Explain why every analysis titration should be repeated at least three times and the result averaged.

9 You are going to titrate $25.0 \, mL$ of $0.205 \, mol \, L^{-1}$ ethanoic acid solution pipetted into a beaker against $0.246 \, mol \, L^{-1}$ potassium hydroxide solution from a burette.

a With what should you rinse the pipette immediately before using it to deliver the $0.205 \, mol \, L^{-1}$ ethanoic acid solution? Why?

b With what should you rinse the beaker before pipetting into it the $0.205 \, mol \, L^{-1}$ ethanoic acid solution? Why?

c With what should you rinse the burette immediately before filling it with the $0.246 \, mol \, L^{-1}$ potassium hydroxide solution ready for titration?

10 For each of the following compounds:

   i list any features of the molecule or functional group(s) that give rise to the properties of the organic compound

   ii identify the class of organic compound

  iii name the compound.

a	$CH_3-CH-CH_3$
	        $\|$
	       OH
		i _____
		
		 ii _____
		
		 iii _____

b	$CH_3-CH-CH=CH_2$
	      $\|$
	     CH_3
		i _____
		
		 ii _____
		
		 iii _____

c	$CH_3-CH_2-CH-C{\overset{O}{\underset{OH}{}}}$
	           $\|$
	         CH_2
	         $\|$
	        CH_3
		i _____
		
		 ii _____
		
		 iii _____

d	$CH_3-CH_2-COOCH_3$	i _____ ii _____ iii _____
e	CH_3 $\|$ $H_3C\diagdown\underset{\underset{H_2}{C}}{{}}\diagup\overset{C}{\underset{H}{}}\diagdown\underset{\underset{CH_3}{\|}}{\overset{H}{C}}\diagup CH_3$	i _____ ii _____ iii _____

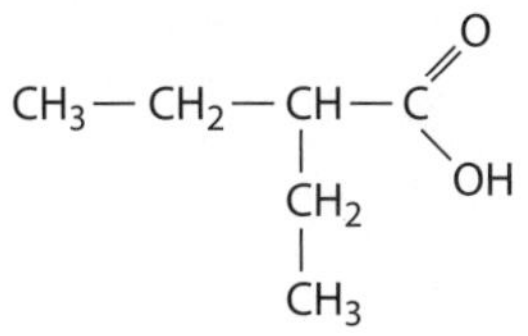

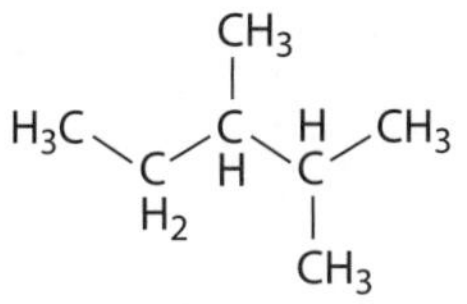

11 Write balanced equations for the following reactions and give the name of the organic product.

   a Pent-1-ene with Cl_2

   b Hex-3-ene with bromine water

   c Methanoic acid and sodium carbonate

d Ethanoic acid and propan-1-ol

12 a Draw the structures of the molecules of a:

    i primary alcohol

    ii secondary alcohol

    iii tertiary alcohol.

b Write the systematic names of the alcohols beside their structures in part a.

c Compare the reactivity and oxidation products from the three alcohols that you nominated in part a when acidified dichromate solution is added to samples of each.

_____

_____

_____

13 A 3.40 g sample of calcium carbonate reacts with 100 mL of 1.00 mol L^{-1} sulfuric acid.

    a Which is the limiting reagent?

_____

_____

_____

_____

_____

    b How much of the excess reagent remains after reaction?

_____

    c What mass of carbon dioxide gas is formed?

_____

_____

INQUIRY QUESTION: HOW ARE THE IONS PRESENT IN THE ENVIRONMENT IDENTIFIED AND MEASURED?

WS 14.1 **Investigating precipitation**

STUDENT BOOK
Pages 407–8

LEARNING GOALS

Evaluate the results of a precipitation investigation

Design an investigation to identify an unknown species

Write balanced net ionic chemical equations

1 A group of students conducted an investigation to observe examples of precipitation reactions. They compared the results of their investigation with a table of solubilities and found that some of the results did not match the precipitation data in the solubility table.

Their results are given below; ✓ means there was a precipitate and × means there was no precipitate.

	Cl^-	SO_4^{2-}	CO_3^{2-}	OH^-
Ca^{2+}	✓	✓	✓	✓
Mg^{2+}	✓	×	✓	✓
Cu^{2+}	×	×	✓	✓
Ba^{2+}	✓	✓	✓	×
Ag^+	✓	✓	✓	✓

The students repeated the testing to check they had not made a mistake and obtained the same results.

a Evaluate the group's data given above and identify any results that are inconsistent with the solubility data table.

b Suggest what could have caused the experimental test result and explain how you came to that conclusion.

2 a Design an investigation (give Aim, Equipment and Method) to determine if your answer to question 1b is correct.

   b Explain how the results of your investigation will determine if your answer is correct.

_____

_____

_____

_____

_____

_____

_____

_____

_____

_____

_____

_____

_____

_____

_____

_____

_____

_____

3 Write balanced net ionic equations for the reactions you used in your investigation in question 2.

_____

_____

_____

Interpret and analyse a flow chart

Evaluate data to identify unknown ions

Write different types of balanced equations

Design a method to separate a mixture of ions

Apply understanding of equilibrium to predict precipitate formation

1 Students were provided with the following flow chart to assist them in identifying cations in solution. Use the flow chart to answer the questions which follow.

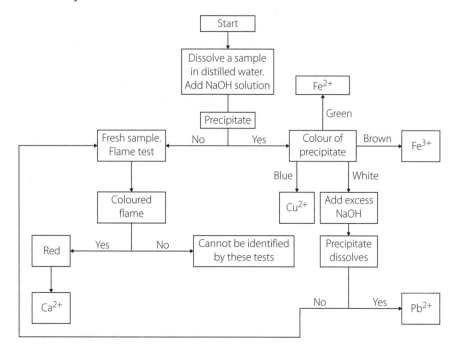

a Which ions are able to be identified using the processes in the flow chart?

_____

b Identify what processes are being used to identify the ions.

_____

c What property must all the cation samples have in order to use the flow chart?

_____

d Explain why distilled water should be used when preparing the samples for testing.

_____

e i A student was given a sample that they were told was one of the ions in the flow chart. They conducted tests and obtained the following results.

A white precipitate was formed when NaOH solution was added to the sample. Upon addition of more NaOH solution, the precipitate dissolved. Which cation was present?

_____

ii When more NaOH is added, the precipitate dissolves. With reference to the equation given below, explain why the precipitate dissolves.

$$Pb(OH)_2(s) + 2OH^-(aq) \rightleftharpoons [Pb(OH)_4]^{2-}(aq)$$

_____

_____

2 Four unknown solutions A–D each contained one of the following anions: CO_3^{2-}, Cl^-, PO_4^{3-}, SO_4^{2-}, CH_3COO^- or OH^-. The solutions were tested to determine which anions were present. The results are given below; ✓ means there was a precipitate and ✗ means there was no precipitate.

Reagent added	A	B	C	D
Ag^+	✓	✓	✓	✓
Ba^{2+}	✗	✗	✗	✓
Pb^{2+}	✓	✗	✓	✓
Mg^{2+}	✓	✗	✗	✗

a Use the results to identify possible anions A–D.

_____

b Write balanced net ionic equations for the reaction between:

 i Ag^+ and D

_____

 ii Pb^{2+} and A.

_____

c If the four anion solutions were mixed together, explain whether or not it would be possible to separate some/ all of the anions. If it is possible, suggest a procedure that could be followed.

_____

_____

_____

_____

_____

_____

_____

_____

_____

_____

3 Testing for phosphates requires the phosphate solution to be alkaline for precipitates to form.

   a Explain why the solution must be alkaline, with reference to the following equation.

$$HPO_4^{2-}(aq) + H_2O(l) \rightleftharpoons PO_4^{3-}(aq) + H_3O^+(aq) \qquad\qquad K_a = 4.8 \times 10^{-13}$$

   b Phosphate ions will precipitate with cations of barium, lead, silver and copper. The K_{sp} of the precipitates of each of these cations is given below.

Cation	Ba^{2+}	Pb^{2+}	Ag^{2+}	Cu^{2+}
K_{sp}	1.3×10^{-29}	8.0×10^{-43}	8.89×10^{-17}	1.40×10^{-37}

    Explain which cation would be the best to use when testing for phosphate ions.

   c What is the minimum concentration of phosphate ions that would precipitate with the cation you identified in part b?

   d What would be the minimum pH of the solution for this precipitation to occur?

   e Comment on the pH needed for precipitates of other cations to form.

Evaluate an experimental design

Process experimental data

Explain differences between experimental and theoretical results

Apply gravimetric analysis techniques to calculate the impurity in a sample

1 Students were told their depth study had to involve the analysis of a substance or product. They were also told that while equipment and chemicals would be provided by the school, they were to provide samples of the substance/product to be analysed. Part of their assessment would consider the efficiency of the analysis and they would be penalised for wastage of chemicals.

One student, having seen a recent media report on the high salt content of soy sauce, decided to analyse a regular and low salt version and compare their results with information provided on the label by the manufacturer. After doing background research and checking labels, they found the following information.

Brand	Type of soy sauce	Sodium content (mg/100 mL)
X	Regular	6833
X	Low salt	3560

Recommended daily salt intake is 5 g salt or 2000 mg sodium.

1 tablespoon (17.5 g) of salt is 291% of the recommended daily intake.

a Suggest why the student chose to use two versions of the same brand.

_____

_____

_____

_____

After further research and reflecting on analysis techniques they had learnt at school, the student decided to analyse the samples using a precipitation titration and identified Volhard's method as the best method to use for their analysis.

They recognised that as soy sauce is dark in colour, it needed to be diluted so the titration colour change could be seen. After doing a number of trials, the student decided that diluting the original sample with distilled water by a factor of 100 would produce a sample light enough for the titration colour change to be identified. For each sample (regular and low salt) the student pipetted 5 mL of the original sample into separate volumetric flasks and topped each of the flasks up to the 500 mL mark with distilled water.

b i In deciding to use a precipitation titration, what ion would the student be measuring and what assumption(s) would they be making?

_____

_____

_____

_____

_____

_____

ii Suggest why they chose Volhard's method rather than the simpler Mohr's method.

_____

_____

_____

_____

_____

_____

_____

_____

_____

_____

_____

c Why did the student use a pipette, volumetric flask and distilled water to dilute the samples?

_____

_____

After doing some calculations based on their research, the student requested the following chemicals:

▶ 25 mL of $1.00 \, mol \, L^{-1}$ AgNO$_3$
▶ 500 mL of $0.020 \, mol \, L^{-1}$ NaSCN
▶ 5 mL of $1.0 \, mol \, L^{-1}$ Fe(NO$_3$)$_3$

They then proceeded to do the following:

▶ Measure out 200 mL of each diluted soy sauce sample into separate beakers.
▶ Add 10.0 mL of AgNO$_3$ to each beaker.
▶ Allow the precipitate to settle, then filter it off.
▶ Add a few drops of Fe(NO$_3$)$_3$ to each of the filtrates.
▶ Measure $4 \times 20.0 \, mL$ aliquots of each filtrate into separate flasks.
▶ Titrate each flask with $0.020 \, mol \, L^{-1}$ NaSCN until a consistent value is obtained.

The student recorded the following results.

Sample	Trial 1 (mL)	Trial 2 (mL)	Trial 3 (mL)	Trial 4 (mL)	Average (mL)
Regular soy sauce	26.5	26.1	25.9	25.9	
Low-salt soy sauce	34.9	34.1	33.9	33.8	

Observation: The precipitate has a brownish colour.

d Calculate the average titre value for each sample and insert the results into the table above.

e i Calculate the number of moles of Ag^+ that reacted in the 20.0 mL samples used in the titration.

_____

_____

_____

ii Calculate the number of moles of excess Ag^+ from the diluted samples of each type of soy sauce.

_____

_____

_____

iii Calculate the number of moles of Ag^+ that reacted with Cl^- in the diluted samples and the number of moles of Cl^- for each type of soy sauce.

_____

_____

_____

_____

_____

iv Calculate the mass of Na^+ per mL in the original undiluted sample of soy sauce.

_____

_____

_____

_____

_____

_____

_____

f Compare the experimental values obtained with those provided by the manufacturer and suggest reasons for any difference.

_____

_____

_____

_____

_____

_____

_____

_____

g Another student in the class asked why this student didn't just use gravimetric analysis as it is much simpler. Suggest a valid response to this question.

_____

_____

_____

_____

_____

2 A sample of aluminium sulfate, $Al_2(SO_4)_3$, was known to be contaminated with aluminium nitrate. The following gravimetric analysis was carried out on a sample of the mixture to determine the extent of the contamination.

A 3.00 g impure sample was dissolved in water. A solution of barium nitrate was added in excess to precipitate the sulfate ions as barium sulfate. The solution was filtered and the precipitate dried. The mass of the dried precipitate was 5.82 g.

a Calculate the number of moles of sulfate ions.

_____

_____

_____

b Calculate the mass of aluminium sulfate.

_____

_____

_____

c Calculate the percentage by mass of the impurity.

_____

_____

WS 14.4 Applying instrumental analysis techniques

LEARNING GOALS

Construct calibration curves

Interpret data from calibration curves

Evaluate the validity of results

1 There was concern expressed that a range of children's toys sold over Christmas had been painted with lead-based paint. Government authorities obtained a range of the toys and used atomic absorption spectroscope to analyse samples of paint from different toys.

Lead samples of known concentration were made and analysed in order to produce a calibration curve. The data from the analysis is given in the table below.

Lead concentration (ppm)	Absorbance
60	4.97
80	6.62
100	8.30
120	9.94

a Construct a calibration curve using the data in the table.

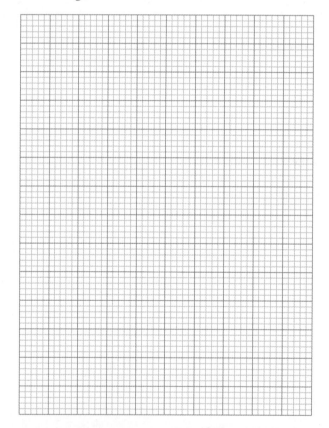

b The following results are the absorbance values for different paint colours.

Sample	Absorbance	Lead concentration (ppm)
1 Yellow	10.25	
2 White	4.50	
3 Red	8.90	
4 Blue	7.85	

For each of the colour samples, plot the absorbance values on the calibration curve and determine the lead concentration of each sample.

c If the maximum allowable limits for lead in paint is $90\,\mathrm{mg\,L^{-1}}$, which if any, of the above samples would be allowable?

_____

_____

d Comment on the validity of the conclusions drawn from this graph.

_____

_____

_____

_____

_____

_____

2 The production manager of a nickel smelter was responsible for ensuring the smelter was operating as efficiently as possible as well as making sure waste discharged from the plant did not contaminate the local waterways.

One way of checking both these aspects was to conduct regular checks of the waste discharge to monitor nickel levels and also those of other contaminants. There was no discharge level mandated for nickel.

The following calibration curve was used to check nickel levels in waste discharged.

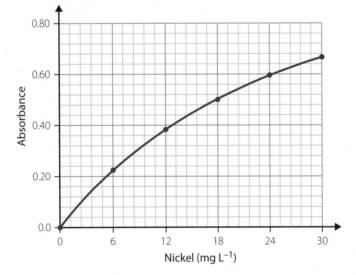

The regular absorbance values for nickel in the discharge ranged between 0.50 and 0.60.

a If the quantity of waste discharge was, on average $1312\,\mathrm{L}$ per day, how much nickel would be lost in a year? (Assume the smelter operated 365 days.)

_____

_____

_____

_____

b A more efficient process developed by CSIRO was implemented. After this process was established, the average absorbance values for nickel dropped to 0.20. How much less nickel was introduced into the waste stream?

_____

_____

_____

3 In the human body, iron is an essential element. It is found in haemoglobin and is responsible for binding oxygen. Iron deficiency is a serious problem, especially in women, and iron supplements are commonly used.

The concentration of iron in the blood can be determined through the process of colourimetry. The iron in a blood sample can be determined by reacting the sample with a suitable reagent, then comparing the absorbance value with a standard calibration curve.

Standard samples were prepared and a suitable filter was used to obtain the following absorbance values.

Concentration Fe^{2+} (mg L^{-1})	Absorbance (%)
2.0	10
4.0	20
8.0	40
12.0	60
16.0	80
20.0	100

a Construct a calibration curve using the values in the table.

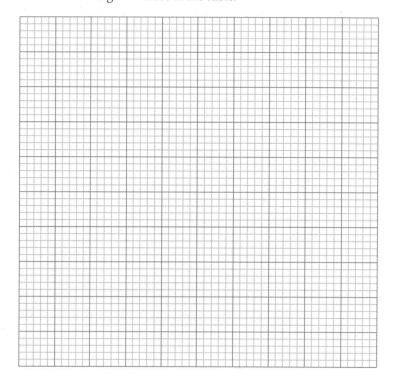

b The blood of a patient thought to have low levels of iron was diluted by a factor of 10, then analysed and found to have an absorbance of 95%.

   i What was the concentration of iron in the diluted sample?

      _____

   ii What was the actual concentration in the blood?

      _____

   iii If the patient has a blood volume of 5.5 L, what is the total mass of iron in the patient's blood?

      _____

c If a healthy 60 kg adult should contain 2.1 g of iron, explain whether the patient should be concerned.

      _____

      _____

Analysis of organic substances

INQUIRY QUESTION: HOW IS INFORMATION ABOUT THE REACTIVITY AND STRUCTURE OF ORGANIC COMPOUNDS OBTAINED?

WS 15.1 Investigating organic compounds

STUDENT BOOK
Pages 448–50

LEARNING GOALS

Predict the products of chemical reactions

Write balanced chemical equations

Analyse experimental data to identify unknowns

1 Five unlabelled bottles of organic liquids were tested to determine the contents of the bottles. The test results are given in the following table.

Liquid	Flammable	Solid sodium carbonate added	Bromine water added	Acidified potassium permanganate added
A	Yes	No reaction	No reaction	Decolourised
B	Yes	No reaction	No reaction	No reaction
C	Yes	No reaction	Decolourised	Decolourised
D	No	Bubbles of gas	No reaction	No reaction
E	Yes	No reaction	No reaction	Decolourised

a Based on the results in the table, identify the class of organic compound each sample could belong to.

_____

_____

b Suggest a way to determine whether A and E are the same or different chemicals.

_____

_____

_____

_____

2 Compound A, common in oils used for aromatherapy, is shown below.

$$CH_3-CH_2 \quad \quad CH_2-OH$$
$$C=C$$
$$H \quad \quad H$$

a Name the functional groups present in compound A.

_____

b Compound A is reacted with bromine water.

    i What observation would confirm a reaction had taken place?

    _____

    ii Write the equation for the reaction and name the product of the reaction.

c **i** Students were asked to suggest a test that could be used to confirm the presence of the alcohol group. Student M suggested adding drops of acidified potassium permanganate to the sample, while student N suggested refluxing with a small amount of glacial acetic acid and a few drops of concentrated sulfuric acid, in a hot water bath for 10 minutes, then pouring into cold water.

    Comment on the effectiveness of the two suggestions in identifying the presence of the alcohol group.

    _____

    _____

    _____

    _____

    _____

    ii Identify any new functional groups that would occur as the result of both students' suggestion.

    _____

    _____

    _____

    _____

LEARNING GOALS

Explain aspects of a mass spectrometer

Identify unknowns using mass spectra

Apply mass spectroscopy to identify contamination of a sample

1   A student was learning about mass spectroscopy and read the following.

'The firing of high-energy electrons into an organic molecule can cause the initial molecular cation to fragment into a smaller cation and a free radical. The free radical is neutral and uncharged so is not detected by mass spectroscopy.'

Explain why the free radical cannot be detected.

_____

_____

_____

_____

_____

2   The mass spectrum of an unknown alkane is shown below.

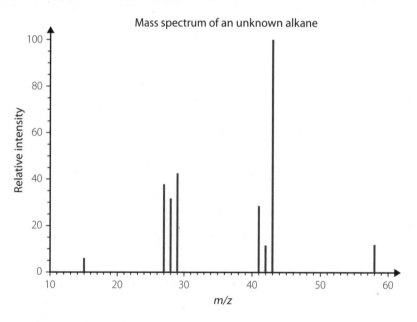

Mass spectrum of an unknown alkane

a   i   Identify the peak for the parent molecular ion.

_____

ii   Determine the molecular formula of the alkane.

_____

_____

b **i** Draw all possible isomers of this alkane.

   ii Name the isomers.

c The unknown alkane was produced by a reaction between an alkene and hydrogen in the presence of a metal catalyst. The original alkene had two possible isomers. Use this information to identify which of your answers to part b is the unknown alkane. Explain.

_____

_____

_____

_____

d Referring to your answer to part c, account for the peaks at 43, 42, 41, 29, 28, 27 and 15 and identify the base peak.

_____

_____

_____

_____

_____

_____

3 Methyl ethanoate is widely used as a solvent in paints, glues and nail polish. It is also an intermediate feedstock for the production of a variety of polymers. A chemical company that produced one of the polymers identified a problem with the methyl ethanoate feedstock during a routine quality assurance check. They produced the following mass spectrum of a sample of the methyl ethanoate used.

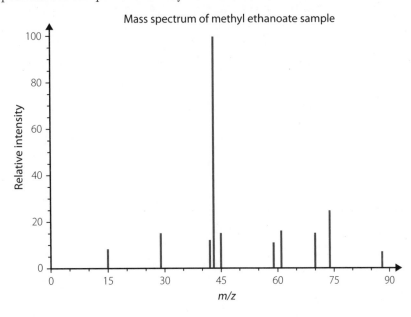

Mass spectrum of methyl ethanoate sample

This was then compared to the reference mass spectrum of methyl ethanoate shown below.

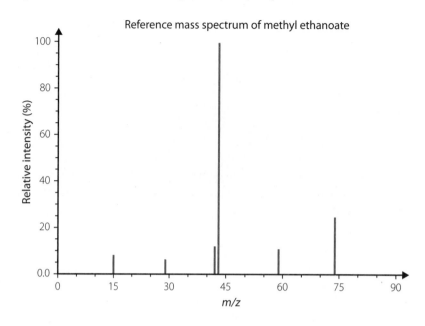

Reference mass spectrum of methyl ethanoate

a Explain whether or not the methyl ethanoate used to manufacture the polymer was contaminated. Use data to support your answer.

_____

_____

_____

b The manufacturer of the methyl ethanoate was contacted. Upon undertaking their own analysis, they agreed the sample was contaminated and undertook to identify the source of the contamination. In their investigation, they identified that the contaminant was either ethyl ethanoate or methyl propanoate. Upon consultation of data sources, the following information was obtained.

Compound	Major peaks (relative intensity ≥ 15)					
Methyl propanoate	88	59	57	29	27	15
Ethyl ethanoate	70	61	45	43	29	-

Identify the contaminant and justify your choice.

_____

_____

_____

_____

9780170449656

Applying NMR spectroscopy techniques

STUDENT BOOK
Pages 461–7

1 Explain the differences between ^1H NMR spectroscopy and ^{13}C NMR spectroscopy.

_____

_____

_____

_____

_____

_____

2 The two ^{13}C NMR spectra of hex-1-ene and hex-3-ene are shown below.

Spectrum A

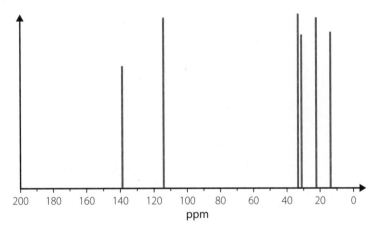

Spectrum B

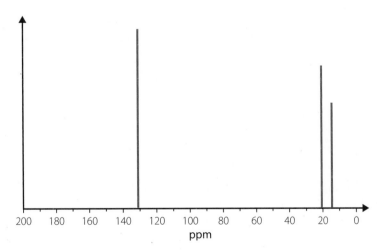

a Draw the structures of hex-1-ene and hex-3-ene.

b Use data to identify which spectrum is of hex-1-ene and which spectrum is of hex-3-ene.

_____

_____

_____

_____

c **i** Predict the total number peaks in a ^{1}H NMR spectrum of hex-3-ene and explain why.

_____

_____

_____

_____

ii Explain what splitting each of these hydrogen environments would show.

_____

_____

_____

_____

_____

_____

3 A sample of compound X was heated with concentrated sulfuric acid to form compound Y. Compound Y reacted with HCl to form compound Z.

a **i** Compound X was analysed and found to be 53.2% carbon, 13.1% hydrogen and oxygen.
Determine the empirical formula of the compound.

_____

_____

_____

_____

_____

ii The molecular weight of the compound is 46.1. Determine the molecular formula.

_____

_____

b The NMR spectra of compounds X, Y and Z are given below. Identify compounds X, Y and Z and explain the reactions described above. For each spectrum, identify peaks used to verify the identity of each compound.

^{1}H NMR spectrum for compound X

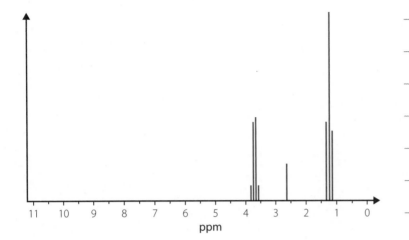

^{13}C NMR for compound Y

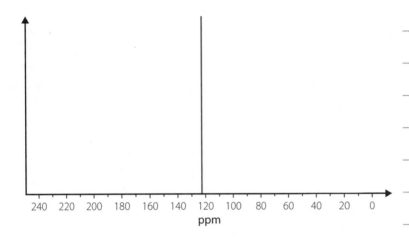

^{1}H NMR spectrum for compound Z

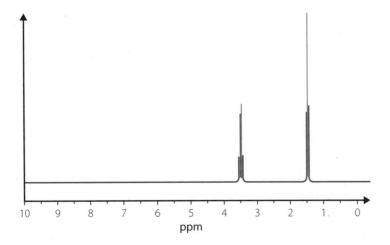

Applying IR and UV–visible spectroscopy techniques

LEARNING GOALS

Compare IR and UV–visible spectra

Match IR spectra to specific compounds

Identify λ_{max} and its use in UV–visible spectroscopy

Construct a calibration curve and interpret data

1 Compare IR and UV–visible spectroscopy techniques.

_____

_____

_____

_____

_____

_____

_____

_____

2 Students conducted an investigation into the oxidation of alcohols. They were provided with propan-1-ol and propan-2-ol. They added acidified potassium dichromate to samples of each of the two alcohols and analysed the product of each reaction using IR spectroscopy. The spectrum of each product is shown below.

Product A

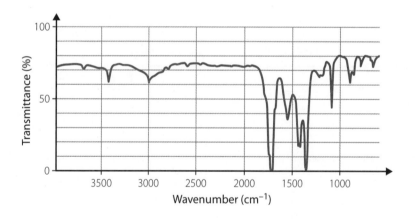

Product B

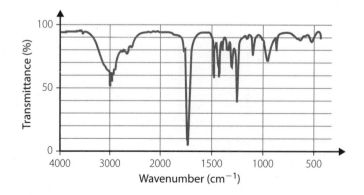

a Predict the products of each of the reactions.

_____

_____

b Identify which of the above spectra matches each of the predicted products. Justify your decision.

_____

_____

_____

_____

_____

_____

_____

_____

c Analyse whether each of the reactions went to completion.

_____

_____

_____

3 As tomatoes ripen, they turn from green due to the chlorophyll pigment to red due to the presence of a hydrocarbon called lycopene ($C_{40}H_{56}$). The absorption spectra for chlorophyll a and lycopene is shown below.

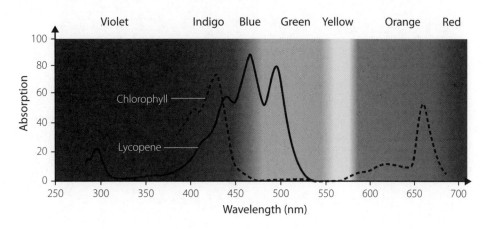

a **i** What is meant by λ_{max}?

_____

ii What is the λ_{max} for lycopene?

_____

b A scientist was researching the change in lycopene concentration as tomatoes ripen. They decided to use UV–visible spectroscopy at a wavelength of 470 nm. Explain why.

_____

_____

_____

_____

4 Carmine is a red dye food colouring. It is found in various cosmetics, drinks and lollies. The acceptable daily intake is 5 mg kg^{-1}. The amount of carmine in a red drink was determined using UV–visible spectroscopy.

Carmine has a λ_{max} at 513 nm. A series of standard dilutions of a stock sample were made and their absorbance measured. The data obtained is given in the table below.

Concentration (ppm)	2	4	6	8	10
Absorbance	0.11	0.23	0.33	0.45	0.51

a Using the data in the table, construct a calibration curve.

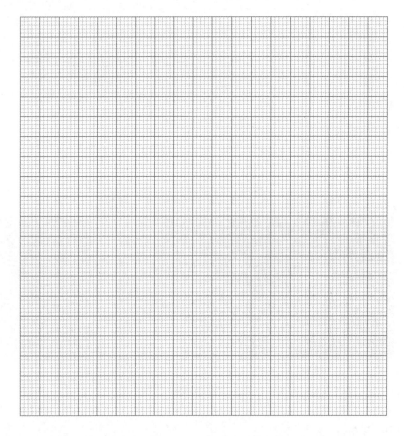

b A 100 mL sample of the red drink to be analysed was diluted by a factor of 10. The diluted sample was analysed and found to have an absorbance of 0.30. Use the calibration curve to determine the concentration of carmine in the sample.

_____

_____

c If a person consumed 500 mL of this drink, how much would they have consumed above the recommended daily limit?

_____

_____

Analyse data to identify unknown samples

Predict spectral identifiers based on molecular structure

Explain and justify identification techniques

1 An analytical chemist was provided with a sample of a compound A. The compound consists of an unbranched carbon chain. Upon analysing the composition of the compound, the chemist found it to have the following percentage composition by mass: C 70.59%; H 13.72%; with the remainder being oxygen. A mass spectrum of the compound showed a molecular ion peak at $m/z = 102.0$. An IR spectrum for compound A is shown below.

IR spectrum of A

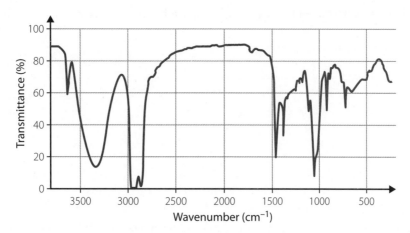

a Use all the information provided to identify all possible structures for A.

_____

_____

_____

_____

_____

_____

_____

_____

_____

_____

_____

_____

_____

_____

_____

b **i** For each of the isomers identified in part a, give the number of different peaks in the ^{1}H NMR and the ^{13}C NMR.

_____

_____

_____

ii Explain whether NMR spectroscopy would be useful for distinguishing between the isomers.

_____

_____

_____

c Suggest another method that could be used to distinguish between the isomers.

_____

_____

_____

_____

_____

_____

2 Justify which of the following ^{13}C NMR, IR and mass spectra belong to each of these compounds: 1,1 dichloropropene, propanone and propan-1-ol.

Mass spectrum A

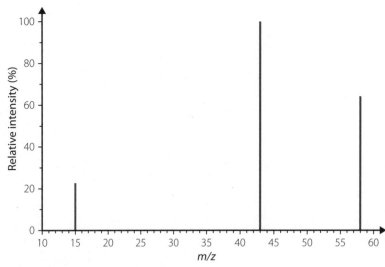

Mass spectrum B

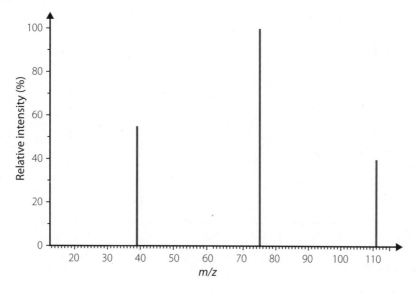

Mass spectrum C

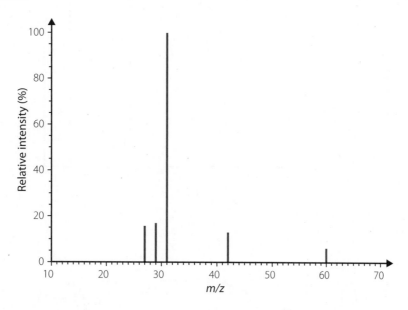

¹³C NMR spectrum D

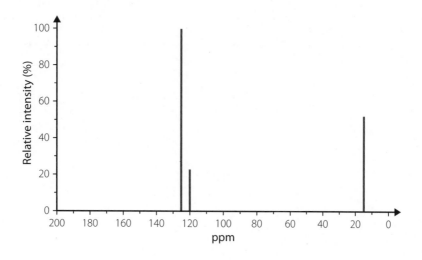

¹³C NMR spectrum E

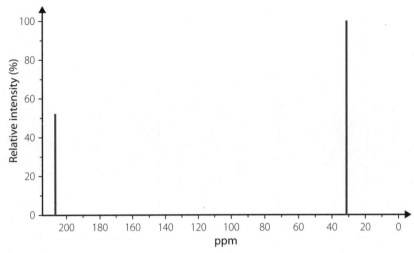

¹³C NMR spectrum F

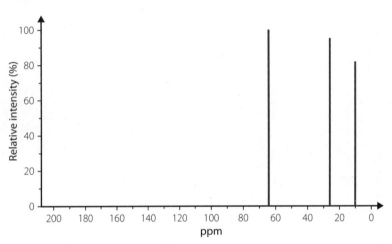

IR spectrum G

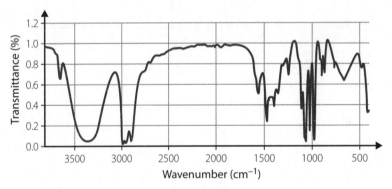

IR spectrum H

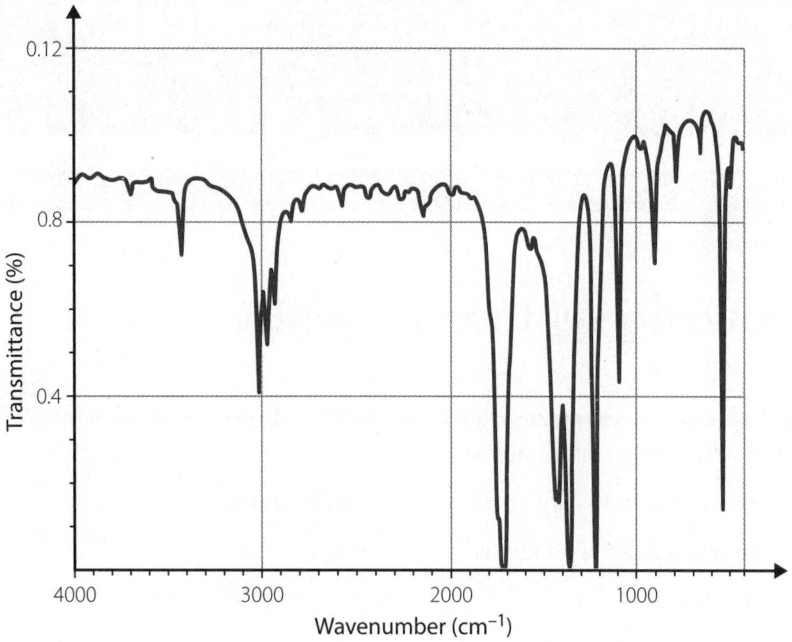

0.12
0.8
0.4
Transmittance (%)
4000
3000
2000
1000
Wavenumber (cm−1)

16 Chemical synthesis and design

WS 16.1 Examining synthesis reactions

STUDENT BOOK
Pages 488–92

LEARNING GOALS

Construct a flow chart based on given information

Identify components and reasons for specific conditions in a synthesis reaction

Analyse the effect of changing reaction conditions

Construct an energy profile diagram

1 Titanium and its alloys have a wide range of uses and are extremely important in many industries as they are strong, light and resistant to corrosion. Most titanium is obtained by processing ores of titanium compounds, mainly rutile (TiO_2) and ilmenite ($FeTiO_3$).

The steps used in the processing of rutile are outlined below.

Step 1 Chlorination

The dry rutile, chlorine gas and carbon (in the form of coke) are mixed, then heated to a temperature of 1300 K which is maintained by the exothermic reaction. The products of the reaction are $TiCl_4(g)$ and $CO_2(g)$.

Step 2 Purification

The impure titanium(IV) chloride is treated with hydrogen sulfide to remove $VOCl_3$ impurity, and then distilled. The titanium(IV) chloride liquid product collected following distillation is >99.9% pure. This product may be used directly to make catalysts and electronics or further oxidised or reduced.

Step 3 Oxidation

The titanium(IV) chloride is a volatile liquid that is readily oxidised by $O_2(g)$ to produce $Cl_2(g)$ and $TiO_2(s)$. The chlorine produced is recycled to be used in Step 1. The titanium(IV) oxide is used to make a range of products, including paints, cosmetics and plastics.

Step 3 Reduction

Most titanium metal is produced using the Kroll process. In this process, titanium(IV) chloride is heated and the vapour is passed into a special reactor, which contains molten magnesium preheated to 800 K in an atmosphere of argon. The reduction of chlorides occurs slowly so the temperature is raised to 1300 K to complete the reduction process. After 36–50 hours, the reactor is allowed to cool, which takes approximately 4–5 days. The magnesium chloride product undergoes electrolysis after it has been separated from the titanium metal. Both the magnesium and chlorine are recycled back into the production process. The titanium metal produced is purified, and then processed into ingots. The ingots are mainly used to make titanium alloys.

a Use the information provided in the description above to draw a flow chart outlining the process of titanium ore. Include relevant chemical equations.

b Refer to your flow chart and identify the following.

   i Raw materials

_____

   ii Final products

_____

   iii Recycled chemicals

_____

   iv Unused or waste products

_____

c **i** Explain why the reduction of chlorides occurs slowly.

_____

_____

_____

_____

HINT

Think about the stability of the chloride ion.

ii Explain why increasing the temperature was necessary for the reduction step.

_____

_____

_____

_____

iii Given Mg has M.P. = 650°C and B.P. = 1110°C, suggest why the temperature was not raised even higher.

HINT

Consider the states of the reactants and where the reaction will be occurring.

_____

_____

_____

_____

2 Benzene is an important chemical widely used as a feedstock to produce other chemicals. Two chemicals that are manufactured from benzene are propanone and phenol. The production occurs through a two-step process shown below.

Step 1

Benzene + Propene ⇌ Isopropylbenzene

Step 2

Isopropylbenzene + O_2 → Propanone + Phenol

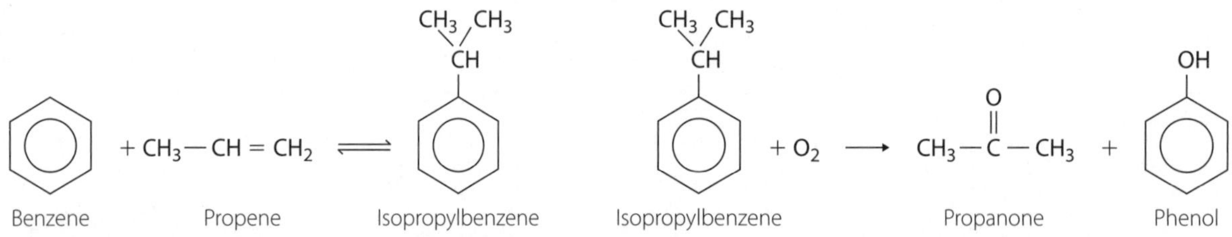

The melting and boiling points of the compounds are given below.

Compound	Benzene	Propene	Isopropylbenzene	Propanone	Phenol
MP (°C)	5.5	−185	−96	−95	40.9
BP (°C)	80.1	−47.7	152	56	181

a The reaction conditions for step 1 are 250°C and 30 atm pressure.

i Identify the state of the reactants and products at this temperature.

_____

_____

ii Explain the effect of increasing the pressure on the product yield in step 1.

_____

_____

b **i** The reaction in step 1 is exothermic. Explain whether decreasing temperature would have any effect on the reaction.

_____

_____

_____

ii Suggest why a lower temperature is not used.

_____

_____

_____

_____

c **i** Identify any intermediates in the reaction.

_____

ii Suggest why the reaction is carried out in two steps rather than just one direct step.

_____

_____

_____

_____

iii Draw a possible energy profile diagram for the two-step reaction. Step 2 reaction is also exothermic.

LEARNING GOALS

Compare yield and purity

Calculate yield and percentage yield

Determine the purity of a product

Analyse a claim related to purity

1 a How can the percentage yield of a chemical process be determined?

_____

_____

_____

_____

_____

b What is the difference between percentage yield and purity?

_____

_____

2 Nitrogen is important for the growth of plants, so most fertilisers contain nitrogen-based compounds. These may be in the form of ammonium sulfate or nitrate ions. Students chose to analyse a particular brand of fertiliser to determine its nitrogen content and compare their results with the label.

They selected a fertiliser in which, according to the label, the only nitrogen-based compound was ammonium sulfate.

They followed the process outlined below:

▸ Weigh a 3.00 g sample of fertiliser and make up a 250 mL solution.

▸ Add 25 mL portions to excess NaOH.

▸ Bubble the ammonia gas produced into a 50.00 mL solution of $0.100 \, \text{mol L}^{-1}$ HCl.

▸ Titrate the resultant solution with $0.100 \, \text{mol L}^{-1}$ NaOH.

After conducting many trials, they determined the average titrant volume to be 45.5 mL.

a Calculate the percentage by mass of ammonium sulfate in the fertiliser mixture.

_____

_____

_____

_____

_____

_____

_____

_____

_____

_____

_____

_____

_____

_____

_____

_____

b If the labelling on the packet stated $(NH_4)_2SO_4$ as 24% by mass, what should the theoretical yield for the reaction have been?

_____

c Assuming the labelling on the packet was correct, what percentage yield did the students achieve?

_____

3 Glycerol, $C_3H_5(OH)_3$, is used as a water-soluble lubricant. While most glycerol is obtained from fats, it can also be produced synthetically. A small-scale trial was conducted to produce glycerol from propene (C_3H_6) via a four-step process. If 15 g of propene produced 1 g of glycerol, what is the percentage yield for the process?

_____

_____

_____

4 Calcium carbonate is the active ingredient in many antacid tablets. A sample of calcium carbonate used in the manufacture of antacid tablets was analysed to test its purity. As $CaCO_3$ is insoluble, a back titration was used to determine the purity of the sample.

A 1.32 g sample of $CaCO_3$ was weighed and added to 25.0 mL of 1.50 mol L^{-1} HCl. The mixture was transferred to a 250 mL volumetric flask and made up to the mark with distilled water. 25.0 mL portions of the sample were titrated with 0.0500 mol L^{-1} NaOH and the average titrant was calculated to be 23.2 mL.

Calculate the purity of the sample.

_____

_____

_____

_____

_____

_____

_____

_____

_____

_____

_____

_____

_____

_____

5 For hand sanitisers to be effective in killing a virus, they need to be at least 70% alcohol. The most common alcohols used in hand sanitisers are ethanol and isopropyl alcohol (propan-2-ol).

There was a lot of concern in the community that many of the hand sanitisers being sold did not contain the required minimum amount of alcohol. One particular brand, X, had been highlighted in a media report as containing only 30%. The label on the bottle of Brand X claimed it contained 70% pure alcohol.

Before it could be analysed, the type of alcohol had to be identified. This was done using NMR spectroscopy.

a The two ^{1}H NMR spectra below are those of ethanol and propan-2-ol.

Spectrum A

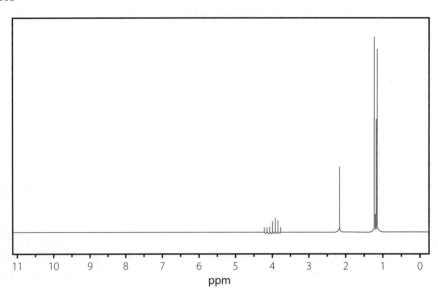

Spectrum B

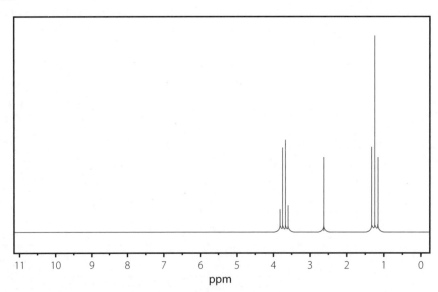

Identify which spectrum is of ethanol and which is of propan-2-ol.

_____

_____

_____

_____

_____

A spectrum of a sample of Brand X matched Spectrum B.

b The concentration of alcohol in a sample of Brand X was determined by redox titration. The following steps were used.

1 The alcohol was oxidised to an aldehyde by warming a diluted sample of sanitiser with an excess solution of acidified potassium dichromate ($K_2Cr_2O_7$).

$$3 \text{ alcohol(aq)} + Cr_2O_7^{2-}(aq) + 8 \text{ H}^+ \rightarrow 3 \text{ aldehyde (aq)} + 2Cr^{3+}(aq) + 7H_2O(l)$$

2 Potassium iodide was added to cause the excess dichromate ions to react to form iodine.

$$Cr_2O_7^{2-}(aq) + 14H^+(aq) + 6I^- \rightarrow 2Cr^{3+}(aq) + 3I_2(aq) + 7H_2O(l)$$

3 This iodine was titrated with sodium thiosulfate, $Na_2S_2O_3$.

$$I_2(aq) + 2S_2O_3^{2-}(aq) \rightarrow 2I^-(aq) + S_4O_6^-(aq)$$

A 10 mL sample of the sanitiser was added to a 250 mL volumetric flask and topped up to the mark with distilled water. 20 mL aliquots were put into conical flasks. To each flask was added 20 mL of 0.40 mol L^{-1} potassium dichromate solution plus 10 mL 40% H_2SO_4. The flasks were then put into a water bath, removed after 10 minutes and had 2 g potassium iodide added. Each of the flasks were titrated against a 1.0 mol L^{-1} solution of sodium thiosulfate using a starch indicator. The titrant value averaged 31.6 mL.

i Calculate the number of moles of $S_2O_3^{2-}$ ions.

_____

ii Calculate the number of moles of I_2 reacted.

_____

iii Calculate the number of moles of $Cr_2O_7^{2-}$ that produced the I_2.

_____

iv Calculate the $n(Cr_2O_7^{2-})$ that reacted with alcohol.

_____

_____

_____

v Calculate the n(alcohol).

_____

vi Calculate the mass of alcohol in the original 10 mL sample.

_____

_____

_____

vii Given the density of the alcohol is 0.785 g mL^{-1}, calculate the volume of alcohol in the 10 mL sample and the percentage alcohol in the sample.

_____

_____

c Comment on the claims made both in the media and by the manufacturer.

_____

_____

_____

Module eight: Checking understanding

1 Hard water contains high concentrations of Ca^{2+} and Mg^{2+} ions. Commercial products called water softeners can be used to remove the Ca^{2+} and Mg^{2+} ions. Which of the following products would be most effective in softening the water?

   A *Enviro soft*, which contains calcium hydroxide and sodium chloride

   B *Ever soft*, which contains magnesium sulfate and sodium chloride

   C *Clear and clean*, which contains potassium sulfate and sodium fluoride

   D *No more scum*, which contains iron iodide and ammonium hydroxide

2 A few drops of dilute copper nitrate solution were added to a solution containing an unknown anion. A blue precipitate formed. The anion most likely present in the solution is which of the following?

   A Cl^-

   B Br^-

   C OH^-

   D SO_4^{2-}

3 Ag^+ forms a bond with NH_3 molecules, resulting in the complex ion $[Ag(NH_3)_2]^+$. Which of the following statements best represents the bonding between the Ag^+ ion and the NH_3 ligand?

   A A hydrogen bond is formed between the Ag^+ ion and a hydrogen of the NH_3 molecule.

   B An ionic bond is formed between the Ag^+ ion and the nitrogen of the NH_3 molecule.

   C A hydrogen of the NH_3 molecule donates an electron to the Ag^+ ion to form a covalent bond.

   D The nitrogen of the NH_3 molecule donates electrons to the Ag^+ ion to form a covalent bond.

4 The diagram below shows the mass spectrum of an organic compound.

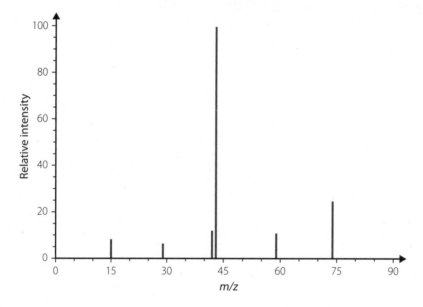

   Which of the following compounds would match the spectrum?

   A Methyl ethanoate

   B Ethyl methanoate

   C Propanoic acid

   D Propanone

9780170449656

5 The ^{1}H NMR spectrum of a molecule that has the molecular formula $C_4H_{10}O$ is shown below.

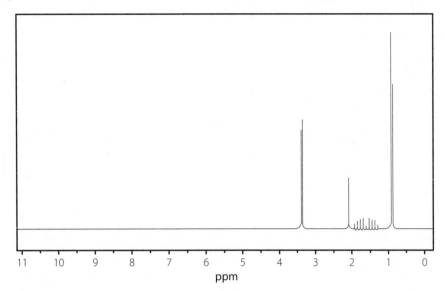

Which of the following is the structural formula of the molecule?

A

$$CH_3 - CH_2 - CH_2 - CH_2$$
$$|$$
$$OH$$

B

$$CH_3$$
$$|$$
$$CH_2 - CH - CH_3$$
$$|$$
$$OH$$

C

$$CH_3$$
$$|$$
$$CH_3 - C - CH_3$$
$$|$$
$$OH$$

D

$$CH_3 - CH_2 - CH - CH_3$$
$$|$$
$$OH$$

6 The concentration of mercury in a 20 mL sample of effluent was determined using atomic absorption spectroscopy. The sample was diluted to 100 mL and the absorbance of the resulting solution was 0.65. Using the calibration curve for mercury below, determine the concentration of mercury in the sample.

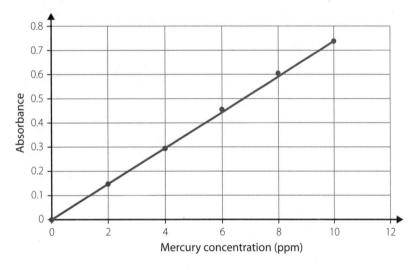

The concentration of mercury in the original sample is

A 0.46 ppm.

B 1.8 ppm.

C 8.8 ppm.

D 44 ppm.

7 Styrene is a monomer that is used in the production of polystyrene. It is mainly produced by catalytic dehydrogenation of ethylbenzene. The amount of styrene produced is affected by the establishment of equilibrium in the reactor. The reaction is endothermic and the industrial process is usually conducted at temperatures between 550 and 650°C.

$$\text{ethylbenzene(g)} \rightleftharpoons \text{styrene(g)} + H_2(g)$$

Which of the following combination of conditions would maximise the yield of styrene?

A Higher temperature, lower pressure

B Higher temperature, higher pressure

C Lower temperature, lower pressure

D Lower temperature, higher pressure

8 Which of the following is the infrared spectrum for propanal?

A

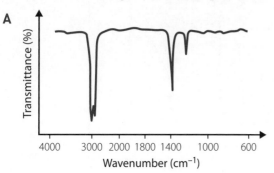

C

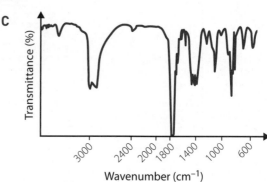

B

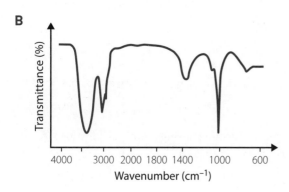

D

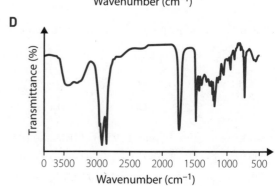

9 A 2.00 g sample of an ore containing lead and copper was dissolved in nitric acid. Sulfuric acid was then added to the solution, which precipitated 0.94 g of $PbSO_4$.

What is the percentage of lead in the sample?

A 25.0%

B 32.1%

C 47%

D 68.7%

10 Students were given a project in which they were asked to determine the percentage of iron in an iron supplement. Which of the following techniques would produce the most accurate analysis?

A Mass spectroscopy

B Infrared spectroscopy

C Atomic absorption spectroscopy

D Gravimetric analysis

11 A solution was thought to contain barium, magnesium and lead ions. Draw a flow chart to show a sequence of steps that could be followed to identify the ions.

12 A student is investigating the purity of a sample of iodine that was being used to add to salt to produce iodised salt. They weighed a 0.80 g sample of the impure iodine, I_2, and dissolved it in a potassium iodide solution.

They conducted a titration using sodium thiosulfate. The titration required 26.2 mL of 0.200 mol L^{-1} sodium thiosulfate to reach end point.

The equation for the reactions is: $I_2(aq) + 2S_2O_3^{2-}(aq) \rightarrow S_4O_6^{2-}(aq) + 2I^-(aq)$

Calculate the percentage purity of the iodine.

13 An organic synthesis was conducted and after reaction an aqueous and organic layer could be seen. The layers were separated, and granules of calcium chloride were added to remove any remaining water from the organic layer. The IR spectrum of the organic layer is shown below.

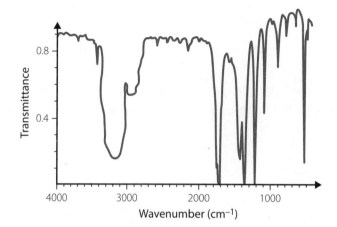

Based on the synthesis reaction, the product is expected to be propanone.

a Explain how the spectrum shows that propanone could be present.

_____

_____

b What other information could be deduced from the spectrum?

_____

_____

_____

14 To determine the solubility product constant (K_{sp}) of PbI_2(s), the solubility of the substance has to be measured at equilibrium. Colourimetry was used to measure the concentration of I^-(aq), which is yellow in solution.

A set of colour standards was made by precisely weighing and dissolving a range of masses of potassium iodide to give concentrations either side of the apparent concentration in the lead iodide saturated solution. These standards are used to produce a calibration curve.

A blue filter is used in the colourimeter.

The following data was obtained from a set of standards of I^-(aq).

Absorbance	0.25	0.49	0.76	1.0	1.24
Conc. I^-(aq) (mol L^{-1}) ($\times 10^{-3}$)	1.0	2.0	3.0	4.0	5.0

The absorbance of a saturated lead iodide sample was 0.68.

a Explain why a blue filter was used in the colourimeter.

_____

_____

_____

_____

b Construct a calibration curve using the data in the table.

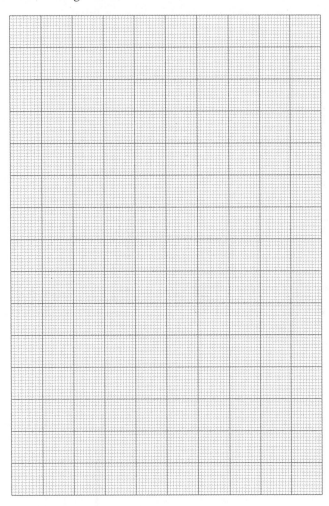

c i Write the K_{sp} expression for lead iodide.

_____

_____

ii Given the absorbance of a saturated lead iodide sample was 0.68, calculate the K_{sp} of lead iodide.

_____

_____

_____

_____

_____

15 Compound A is reacted with water in the presence of a dilute sulfuric acid catalyst to produce compound B.

Compound B is heated with potassium dichromate and sulfuric acid to form compound C.

The mass spectrum of compound A, IR spectra of compounds A, B and C, and the ^{1}H NMR spectra of compounds A and C are given.

Identify compounds A, B and C and explain the reactions described above. For each spectrum, specify the presence or absence of at least two peaks to support your identification.

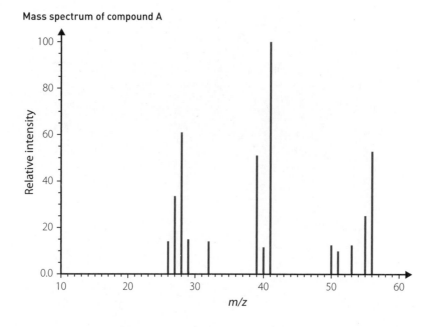

Mass spectrum of compound A

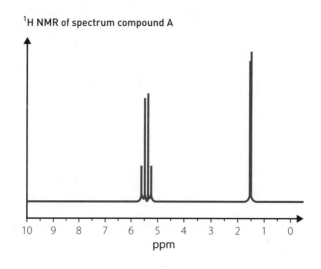

¹H NMR of spectrum compound A

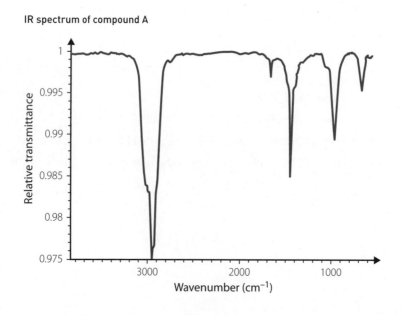

IR spectrum of compound A

IR spectrum of compound B

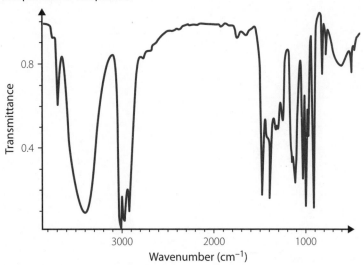

IR spectrum of compound C

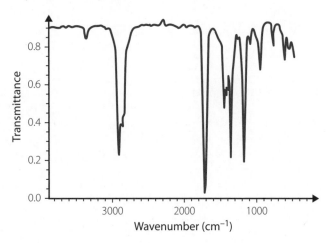

^{1}H NMR spectrum of compound C

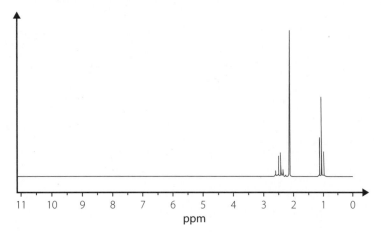

Section I

20 marks

1 What is the IUPAC name for the compound shown?

$$CH_2-CH-CH-CH-CH_3$$

with substituents CH_3, CH_3 on the third and fourth carbons, and CH_2-CH_3 below the first CH.

   A 1-Ethyl-3-4-dimethylpentane

   B 2,3-Dimethylheptane

   C 2,3-Dimethylhexane

   D 5,6-Dimethylheptane

2 A hydrocarbon underwent complete combustion, producing 50 mL of carbon dioxide and 50 mL of water vapour at the same temperature and pressure. What is the most likely formula of the hydrocarbon?

   A C_2H_2

   B C_2H_2O

   C C_2H_4

   D C_2H_4O

3 The production of menadione, a chemical that aids blood clotting, from naphthalene occurs in two steps. Both steps involve the use of a catalyst. The second step is represented by the following equation.

$$\text{2-methylnaphthalene(l)} \underset{}{\overset{CrO_3}{\rightleftharpoons}} \text{menadione(l) + heat}$$

The yield from a particular synthesis resulted in only 0.001 mol of menadione being produced from 0.01 mol of 2-methylnaphthalene.

Which of the following would best explain the poor yield?

   A Decreasing the pressure of the system

   B Heating the reaction mixture

   C Precipitation of menadione as it forms

   D Using excess catalyst CrO_3

4 Propene is reacted with hydrogen chloride. What is the major product?

   A 1-Chloropropane

   B 2-Chloropropane

   C 1,2-Dichloropropane

   D 3-Chloropropane

5 UV–visible spectroscopy was used to monitor the production of an ester in an industrial process. A series of 20 mL samples were extracted from the reaction vessel at regular intervals, then diluted to 100 mL.

The calibration curve used for the ester is shown below.

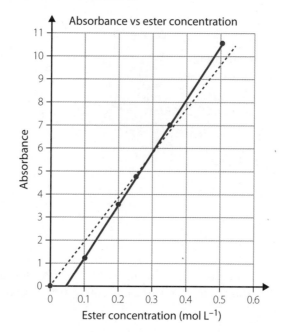

What was the concentration of the ester in the original sample if the absorbance of the measured sample was 4.0?

A $0.044 \, \text{mol L}^{-1}$

B $0.22 \, \text{mol L}^{-1}$

C $1.1 \, \text{mol L}^{-1}$

D $8.0 \, \text{mol L}^{-1}$

6 A 25.0 mL sample of a $0.100 \, \text{mol L}^{-1}$ acetic acid solution completely reacted with 36.4 mL of potassium hydroxide solution. What volume of the same potassium hydroxide solution would be required to completely react with 25.0 mL of a $0.100 \, \text{mol L}^{-1}$ hydrochloric acid solution?

A Less than 36.4 mL

B 36.4 mL

C More than 36.4 mL

D Unable to calculate unless the concentration of the potassium hydroxide solution is also known

7 A stock solution of hydrochloric acid with pH 1.8 was diluted with a volume of water to produce a new solution with pH 4.8.

Which row in the table correctly identifies an appropriate volume of both the original solution and the volume of water added to achieve this dilution?

	Volume of stock solution (mL)	Volume of water added (mL)
A	1	999
B	1	1000
C	10	990
D	10	1000

8 Which acid–base pair could act as a buffer in solution?

A CH_3COOH and CH_3COO^-

B H_2O and OH^-

C HCl and Cl^-

D NaOH and OH^-

9 The ^{13}C NMR spectrum of a molecule with the molecular formula C_3H_6O is shown below.

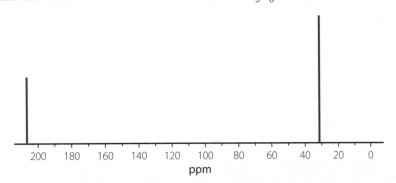

Which of the following is the structural formula that best matches the spectrum?

A

CH_3—CH_2—C
with O double-bonded and H single-bonded

B CH_3—C—CH_3 with O double-bonded below

C CH_2=CH—CH_2—OH

D CH_2—CH=CH_2—OH

10 The graph shows three gases, W, X and Y, that were placed in a sealed container and allowed to reach equilibrium. The temperature was increased at 2 minutes. Another change was made to the system at 5 minutes.

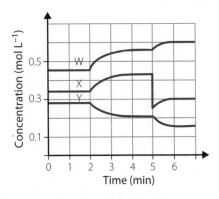

Which alternative correctly identifies the system?

	Equation	Change at 5 minutes
A	$W + X \rightleftharpoons Y$ $\Delta H > 0$	Volume of container was decreased
B	$W + X \rightleftharpoons Y$ $\Delta H > 0$	Volume of container was increased
C	$W + X \rightleftharpoons Y$ $\Delta H < 0$	X was added to the container
D	$W + X \rightleftharpoons Y$ $\Delta H < 0$	X was removed from the container

11 A sample of 1.10 moles of dinitrogen tetroxide was placed into a 2 L vessel and allowed to decompose. At equilibrium, the concentration of nitrogen dioxide was $0.90\,mol\,L^{-1}$. What is the concentration of dinitrogen tetroxide at equilibrium?

A $0.65\,mol\,L^{-1}$

B $0.35\,mol\,L^{-1}$

C $0.20\,mol\,L^{-1}$

D $0.10\,mol\,L^{-1}$

12 The following tests were performed on unknown anions X, Y and Z.

▸ Test 1: Each of the solutions, X, Y and Z, was added to a $0.10\,mol\,L^{-1}$ solution of barium nitrate. X formed a white precipitate. Y and Z did not precipitate.

▸ Test 2: Each of the solutions, X, Y and Z, was added to a $0.10\,mol\,L^{-1}$ solution of magnesium nitrate. Y and Z formed a white precipitate. X did not precipitate.

▸ Test 3: Each of the solutions, X, Y and Z, was tested with red litmus. Y turned the litmus blue. X and Z stayed red.

Which of the following correctly identifies X, Y and Z?

	X	Y	Z
A	Cl^-	OH^-	F^-
B	Cl^-	CO_3^{2-}	PO_4^{3-}
C	SO_4^{2-}	OH^-	F^-
D	SO_4^{2-}	CH_3COO^-	Cl^-

13 An equilibrium was established as shown by the equation. Colours of each species are shown.

$$Fe^{3+}(aq) + SCN^-(aq) \rightleftharpoons FeSCN^{2+}(aq) \quad \Delta H < 0$$
$$\text{pale yellow} \quad \text{colourless} \quad \text{red}$$

A change was imposed and the equilibrium mixture became less red in colour. Which alternative correctly identifies the change?

A $NaOH(aq)$ was added.

B $NH_4SCN(aq)$ was added.

C Pressure was decreased.

D Temperature was decreased.

14 Excess solid AgI is added to 10 mL of each of the following $0.01\,mol\,L^{-1}$ solutions. In which would the *least* amount of AgI dissolve?

A $AgNO_3$

B HI

C NaCl

D PbI_2

15 What is the concentration of barium ions $(mol\,L^{-1})$ in a solution of barium hydroxide with pH 9.3?

A 1.0×10^{-5}

B 2.0×10^{-5}

C 2.5×10^{-10}

D 5.0×10^{-10}

16 The solubility of potassium nitrate is 60 g per 100 g water at 40°C. An 18 g quantity of potassium nitrate is added to a beaker containing 20 g of water at 40°C and stirred until the solution was saturated. What was the mass of undissolved potassium nitrate at 40°C?

A 2 g

B 6 g

C 8 g

D 10 g

17 Barium hydroxide has a K_{sp} value of 2.55×10^{-4}. What is the maximum hydroxide ion concentration possible in a solution with $[Ba^{2+}] = 4.8 \times 10^{-3}\,mol\,L^{-1}$.

A $2.3 \times 10^{-1}\,mol\,L^{-1}$

B $5.3 \times 10^{-2}\,mol\,L^{-1}$

C $9.6 \times 10^{-3}\,mol\,L^{-1}$

D $10.3 \times 10^{-4}\,mol\,L^{-1}$

9780170449656

18 Which alternative shows the correct geometry for each of the labelled carbon atoms?

	Carbon number					
	1	**2**	**3**	**4**	**5**	**6**
A	Linear	Trigonal planar	Trigonal planar	Linear	Linear	Pyramidal
B	Tetrahedral	Tetrahedral	Tetrahedral	Trigonal planar	Pyramidal	Tetrahedral
C	Trigonal planar	Linear	Linear	Tetrahedral	Tetrahedral	Pyramidal
D	Trigonal planar	Linear	Linear	Trigonal planar	Trigonal planar	Tetrahedral

19 The structure of a compound is shown.

Which statement is correct regarding the name of the compound and its production?

	Name	**Production**
A	Ethyl propanone	Oxidation of ethyl propan-2-ol
B	Ethyl propanoate	Refluxing ethanol and propanoic acid
C	Ethyl propanoate	Oxidation of pentan-2-ol
D	Propyl ethanoate	Refluxing propanol and ethanoic acid

20 A sample of sea water was analysed using a precipitation titration to determine the chloride ion concentration. A 25 mL sample of the sea water required 13.7 mL of a $1.00 \, mol \, L^{-1}$ solution of silver nitrate to reach end point.

Which of the following is the value for the chloride ion concentration?

A $0.00055 \, mol \, L^{-1}$

B $0.0137 \, mol \, L^{-1}$

C $0.343 \, mol \, L^{-1}$

D $0.548 \, mol \, L^{-1}$

Section II

80 marks

Question 21 (3 marks)

A solution was made by mixing 42.0 mL of 0.220 mol L^{-1} hydrochloric acid with 55.0 mL of 0.100 mol L^{-1} barium hydroxide. What is the pH of the final solution?

_____

_____

_____

_____

_____

_____

_____

_____

_____

_____

Question 22 (8 marks)

The decomposition of dinitrogen tetroxide to brown nitrogen dioxide is an exothermic reaction shown in the equation below.

$$N_2O_4(g) \rightleftharpoons 2NO_2(g) \quad \Delta H = \text{positive}$$

a Draw a reaction rate versus time graph to show the above reaction reaching equilibrium. (2 marks)

b In terms of collision theory, account for the shapes of the curves you drew in part a. (3 marks)

_____

_____

_____

c The equilibrium mixture was heated. Justify the colour change observed. (3 marks)

_____

_____

_____

_____

_____

Question 23 (6 marks)

a Draw the structure for the monomer(s) that make up the polymer polyethylene terephthalate (PET), shown below.
(2 marks)

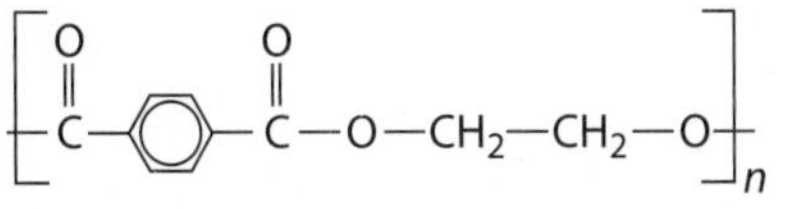

b Name the type of chemical reaction by which this polymer is produced. (1 mark)

_____

c PET is a thermosetting polymer. Use the structure of PET and this fact to account for its uses. (3 marks)

_____

_____

_____

_____

_____

Question 24 (3 marks)

One step in the industrial process used to manufacture sulfuric acid is the exothermic reaction between oxygen and sulfur dioxide to form sulfur trioxide.

Write the equilibrium expression for this reaction and calculate the equilibrium constant at 25°C when $3.21\,mol\,L^{-1}$ oxygen and $4.70\,mol\,L^{-1}$ sulfur dioxide reacted within a 1.0 L vessel. The final concentration of the oxygen gas was $1.35\,mol\,L^{-1}$. (3 marks)

_____

_____

_____

_____

_____

_____

_____

Question 25 (6 marks)

a Account for the fact that two solutions of bases with the same concentration can have different pH values.   (3 marks)

_____

_____

_____

_____

_____

_____

b Analyse how the theories of Brønsted–Lowry and Arrhenius can be applied to sodium hydroxide and
ammonia.   (3 marks)

_____

_____

_____

_____

_____

_____

_____

Question 26 (5 marks)

Aqueous silver ions are involved in the following equilibria:

$$Ag^+(aq) + Cl^-(aq) \rightleftharpoons AgCl(s)$$
$$Ag^+(aq) + 2NH_3(aq) \rightleftharpoons Ag(NH_3)_2{}^+(aq)$$

a If sodium chloride and silver nitrate solutions are mixed, a white precipitate of silver chloride forms. If ammonia is
then added, the precipitate dissolves to form a clear solution. Account for these observations in terms of the above
equilibria.   (2 marks)

_____

_____

_____

_____

_____

b If drops of nitric acid are then added to the clear solution formed in part a, the precipitate re-forms.
Explain why.   (3 marks)

_____

_____

_____

_____

_____

_____

_____

_____

Question 27 (4 marks)

A student titrated acetic acid, CH_3COOH, with $0.5\,mol\,L^{-1}$ sodium hydroxide, obtaining the following titration curve.

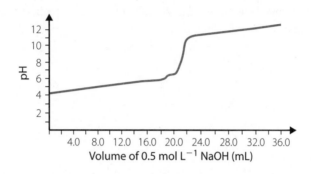

Use the graph to calculate the K_a of the acetic acid used in the titration and analyse the accuracy of the result.

K_a(acetic acid) $= 1.76 \times 10^{-5}$

_____

_____

_____

_____

_____

_____

_____

_____

Question 28 (9 marks)

Sulfuric acid is produced industrially using the contact process. The multistep process for the production occurs through a sequence shown by the equations below.

Step 1: $S(s) + O_2(g) \rightarrow SO_2(g)$

Step 2: $2SO_2(g) + O_2(g) \rightleftharpoons 2SO_3(g)$ $\Delta H = -196\ kJ\,mol^{-1}$

Step 3: $3SO_3(g) + H_2SO_4(l) \rightarrow H_2S_2O_7(l)$

Step 4: $H_2S_2O_7(l) + H_2O(l) \rightarrow 2H_2SO_4(l)$

a Show how to obtain the overall equation for the production of dihydrogen sulfate, $H_2SO_4(l)$, from sulfur dioxide, $SO_2(g)$, by combining the above equations. (2 marks)

b In step 2, the amount of air is controlled so that the ratio of $SO_2 : O_2$ is $1 : 1$. Explain the effect of this on the step 2 process, with reference to efficiency. (4 marks)

c Calculate the yield of dihydrogen sulfate, $H_2SO_4(l)$, that would be expected if the process started with 150 kg of sulfur dioxide, $SO_2(g)$, and the process is 75% efficient. (3 marks)

Question 29 (5 marks)

During the COVID-19 crisis, people were advised to wash their hands with soap. Coronavirus particles are surrounded by a fatty outer layer; hence, soap was found to be very effective at removing the fatty coating surrounding the virus particles.

Using appropriate diagrams, explain the cleaning action of soap and how it is effective in removing the fatty coating on the virus.

_____

_____

_____

_____

_____

_____

_____

_____

_____

_____

Question 30 (5 marks)

The purity of a sample of dry ice (CO_2) was determined using the following procedure.

1 A 3.00 g sample of dry ice was placed in a clean, dry flask.
2 A 100.0 mL sample of $1.50 \, mol \, L^{-1}$ sodium hydroxide was added to the flask to form sodium carbonate and water. The sodium hydroxide was in excess.
3 The flask was sealed and the reaction was allowed to reach completion.
4 The remaining sodium hydroxide was titrated against a $1.50 \, mol \, L^{-1}$ solution of hydrochloric acid. The average volume of hydrochloric acid used was 32.8 mL.

Determine the purity of the sample of dry ice.

_____

_____

_____

_____

_____

Question 31 (7 marks)

A compound was analysed and found to have the empirical formula C_3H_6O. A variety of spectra relating to the compound are shown below. Identify the structure and name the molecule.

Mass spectrum

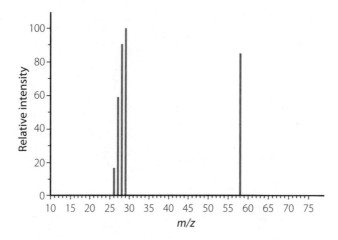

^{13}C NMR spectrum

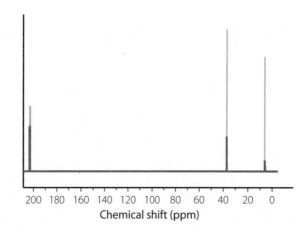

^{1}H NMR spectrum

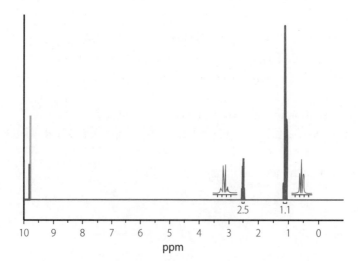

_____

_____

_____

_____

_____

_____

_____

_____

_____

_____

_____

_____

_____

_____

_____

_____

_____

_____

_____

_____

_____

_____

_____

Question 32 (6 marks)

The information provided in the table below shows how temperature affects the solubility product (K_{sp}) of calcium hydroxide.

Temperature (°C)	Solubility product (K_{sp})
10	5.50×10^{-6}
20	4.25×10^{-6}
30	3.21×10^{-6}
40	2.71×10^{-6}
50	2.32×10^{-6}
60	1.91×10^{-6}
70	1.50×10^{-6}

Using the following graph, calculate the solubility of calcium hydroxide in g/100 g water at 35°C. Include a fully labelled graph and relevant chemical equation in your answer.

Question 33 (6 marks)

Monosodium citrate, $NaH_2C_6H_5O_7$, is a salt of citric acid, $H_3C_6H_5O_7$, often used as a buffer in food preservation, an anticoagulant in blood collection and in the medical treatment of kidney stones.

A buffer solution was prepared by mixing $55.0\,mL$ of $0.350\,mol\,L^{-1}$ citric acid solution with $100.0\,mL$ of monosodium citrate solution.

a Calculate the mass of monosodium citrate that would need to be dissolved in $55.0\,mL$ of distilled water to make the most effective buffer solution. (2 marks)

b Explain how this buffer solution can resist pH change upon addition of both acids and bases. (4 marks)

_____

_____

_____

_____

_____

_____

Question 34 (7 marks)

The labels have fallen off four bottles containing colourless liquids. The names on the labels are:

ethanol, ethanoic acid, pentane and pentene.

Outline a method by which all four organic compounds could be identified using appropriate chemical testing. Support your answer with relevant organic structural equations.

_____

_____

_____

_____

_____

_____

_____

_____

_____

_____

_____

_____

ANSWERS

Answers and suggested responses to applicable questions are provided below for you to check your answers. Fully worked solutions are provided for all questions.

MODULE FIVE: EQUILIBRIUM AND ACID REACTIONS

REVIEWING PRIOR KNOWLEDGE PAGE 1

1 a False. An increase in temperature means a decrease in enthalpy.

  b False. Respiration is an exothermic process.

  c True

  d False. The precipitation of silver chloride results in a decrease in entropy because the system becomes more ordered.

  e True

  f False. Gibbs free energy can be calculated at any temperature.

  g False. The particles need to have the minimum energy required for a successful collision and they also need to be in the correct orientation for a reaction to occur.

  h True

  i True

  j False. The activation energy for the forward reaction is always greater in an endothermic reaction than for the reverse reaction because energy must be added for the forward reaction to proceed.

  k False. A catalyst speeds up the rate of a reaction by decreasing the activation energy.

  l True

  m False. Temperature is a measure of the AVERAGE kinetic energy of molecules.

2 a Increase the temperature of the nitric acid; increase surface area by crushing the calcium carbonate into a fine powder; increase the concentration of the nitric acid.

  b $CaCO_3(s) + 2HNO_3(aq) \rightarrow Ca(NO_3)_2(aq) + CO_2(g) + H_2O(l)$

  c

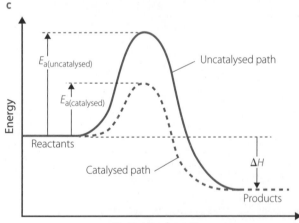

  d i Initially, there is no carbon dioxide because the reaction has not yet started. Then, the volume of carbon dioxide production is rapid, indicating a high rate of reaction. This is because there is the maximum amount of each reacting species at the beginning and the likelihood of successful collisions in the correct orientation with the activation energy is the highest. Then, as the amount of reactants decreases, so does the rate of reaction until it stops because at least one of the reactants has been completely consumed in the reaction, i.e. the point at which the curve becomes horizontal.

ii

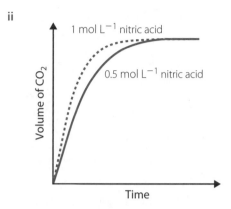

3 a $C_3H_8(g) + 5O_2(g) \rightarrow 3CO_2(g) + 4H_2O(g)$ ΔH is negative

  b There is an increase in entropy because there are more gaseous molecules of products than of reactants, i.e. seven gaseous product molecules versus six gaseous reactant molecules.

4 a $NH_4Cl(s) \rightarrow NH_4^+(aq) + Cl^-(aq)$

  b Ammonium chloride is an ionic compound that is soluble in water and the process of dissolution is endothermic. This means that the energy required to overcome the electrostatic forces of attraction between the ammonium and chloride ions in the ionic lattice is greater than the energy released when ion–dipole bonds form between ammonium ions and the $\delta-$ oxygen atom of water and chloride ions and the $\delta+$ hydrogen atoms of water. Thus, the enthalpy of the system increases. The entropy of the system also increases as the ions are now free to move in the water and the ordered ionic lattice is broken.

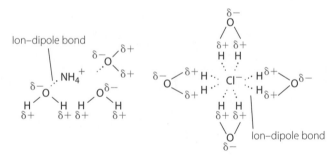

5 The reaction will be spontaneous at 1200 K because ΔG^θ is negative but not at 500 K because ΔG^θ is positive.

6

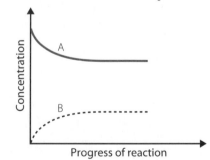

Initially, there is no B as the reaction has not started. Then, the concentration of A gradually decreases and the concentration of B gradually increases as the reaction starts. The reaction eventually stops when there is no change in the concentrations of A or B and the two lines are parallel.

WS 1.1 PAGE 5

1 a 1 Set up the Bunsen burner, tripod and gauze mat.

2 Place about 5 g of the hydrated cobalt(II) chloride hexahydrate in an evaporating basin. Note the colour of the compound.

3 Turn on the Bunsen burner to a blue flame and heat the evaporating basin until the colour of the compound changes. Note the colour.

4 Turn off the Bunsen burner. Using tongs, remove the evaporating basin and allow it to cool.

5 Place the cooled evaporating basin in an ice water bath. Note the colour of the compound.

b

Compound	Observation
Initial	Pink solid
After heating	Pink solid turned blue
After cooling	Blue solid started to turn pink around the edges

c The observed reaction is reversible. Heating the pink solid produced a blue solid and upon cooling, the pink solid re-formed.

d The melting point of the original pink solid and the pink solid formed after cooling can be tested. The melting point should be the same.

2 a $Fe^{3+}(aq) + SCN^-(aq) \rightleftharpoons FeSCN^{2+}(aq)$

b The equilibrium system is an orangey colour, depending on the concentrations of the species involved.

c

Species	Colour
Fe^{3+}	Yellowish
SCN^-	Colourless
$FeSCN^{2+}$	Blood red

d i The equilibrium reaction as written above is exothermic.

$Fe^{3+}(aq) + SCN^-(aq) \rightleftharpoons FeSCN^{2+}(aq) + heat$

yellowish   colourless   blood red

An increase in temperature is causing the equilibrium to shift to the left (the reactant side) because the orange colour is becoming lighter. While cooling, the system is causing the equilibrium to shift to the right (the product side) because the orange colour is becoming darker. According to Le Chatelier's principle, if a system is at equilibrium and a change is made that upsets the equilibrium, then the system alters to counteract the change and a new equilibrium is established. In this case, increasing the temperature favoured the endothermic direction, while cooling favoured the exothermic direction.

ii Sodium hydroxide solution may have been added.

$Fe^{3+}(aq) + 3OH^-(aq) \rightleftharpoons Fe(OH)_3(s)$

This would form the precipitate iron(III) hydroxide, which is a yellow-brown colour and results in a decrease in the concentration of $Fe^{3+}(aq)$, causing the equilibrium to shift to the left (the reactant side), causing the orange colour of the mixture to fade as the concentration of the blood-red iron thiocyanate ion decreases.

iii More potassium thiocyanate solution may have been added.

This would increase the concentration of SCN^- ions, causing the equilibrium to shift to the right (the product side), causing the orange colour to become redder as the concentration of the blood-red iron thiocyanate ion increases.

WS 1.2 PAGE 8

1 a $N_2(g) + 3H_2(g) \rightleftharpoons 2NH_3(g)$ $\Delta H = negative$

b

Before reaction	After reaction

c

	> 0, ~ 0 or < 0
ΔH	< 0
ΔS	< 0
ΔG	~ 0

2 Dynamic equilibrium is represented by flask W because there is nothing happening, and the flask is sealed (it is a closed system). There are no macroscopic changes. The liquid is evaporating by turning into a gas, but the vapours are also condensing into a liquid at the same rate. Static equilibrium is shown by flask X because it is open, and the liquid can evaporate and leave the flask. This process of the liquid turning into a gas and leaving the flask shows a static equilibrium because eventually the flask will be empty.

WS 1.3 PAGE 9

1

	Reaction	Balanced equation	ΔH (+ or −)	ΔS (+ or −)
a	Carbon monoxide gas reacts with oxygen gas to produce carbon dioxide gas. Heat is released.	$2CO(g) + O_2(g) \rightarrow 2CO_2(g)$	−	−
b	Liquid octane, C_8H_{18}, reacts with oxygen gas to produce carbon dioxide and steam. Heat is released.	$C_8H_{18}(l) + 12.5O_2(g) \rightarrow 8CO_2(g) + 9H_2O(g)$	−	+
c	Solid glucose, $C_6H_{12}O_6$, reacts with oxygen to produce carbon dioxide gas and water vapour. Heat is released.	$C_6H_{12}O_6(s) + 6O_2(g) \rightarrow 6CO_2(g) + 6H_2O(g)$	−	+
d	Carbon dioxide and water vapour react in the presence of chlorophyll and sunlight to produce solid glucose and oxygen gas. Heat is absorbed.	$6CO_2(g) + 6H_2O(g) \rightarrow C_6H_{12}O_6(s) + 6O_2(g)$	+	−

2 a $CH_3CH_2OH(l) + 3O_2(g) \rightarrow 2CO_2(g) + 3H_2O(g) + heat$

b The spark or the flame provides the activation energy required for the reaction to take place. This is the minimum energy required for the reaction. Once the reaction is started it produces heat; therefore, enthalpy of the products is less than the enthalpy of the reactants. While entropy increases because there are 3 moles of gaseous particles on the left and 5 moles of gaseous particles on the right, both of these factors contribute to a spontaneous reaction.

3 a $NaNO_3(s) \rightarrow Na^+(aq) + NO_3^-(g)$

b Enthalpy is positive. Entropy is positive.

c Enthalpy favours the reverse reaction, while entropy favours the forward direction. The entropy drive is stronger as the reaction proceeds in the direction written.

4 a
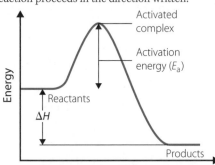

b i The reaction is exothermic because the energy of the products is less than the energy of the reactants.

ii An increase in temperature increases the average kinetic energy of the particles A and B. Therefore, more particles will have collisions with enough energy to overcome the activation energy for the reaction to occur faster, i.e. per unit time, which results in an increase in the reaction rate.

c The activation energy of both the forward and reverse reactions must be low enough that enough particles will have enough energy for successful collisions in either direction for a reversible reaction to occur.

WS 1.4 PAGE 11

1 a $W + X \rightleftharpoons Y$

Initially, the concentration of Y is 0; therefore, Y is not present at the start. Then, as concentrations of W and X decrease, the concentration of Y increases; therefore, W and X must be the reactants and Y the product. The reaction reaches equilibrium at T_1 because after that time the concentrations of all species remain the same, as can be seen by equimolar changes in all species, i.e. Y increases by 1.5 moles, while W and X decrease by 1.5 moles; hence, the mole ratio $X:Y:Z = 1:1:1$.

b Equilibrium lies to the left (the reactant side) because there is a higher concentration of reactants than products.

c
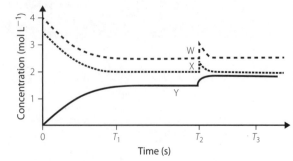

2 a $2SO_2(g) + O_2 \rightleftharpoons 2SO_3(g)$

According to Le Chatelier's principle, a decrease in temperature would favour the reaction that produces heat. Since $[SO_3]$ increases, it indicates that the forward reaction produces heat; hence, the reaction is exothermic.

b
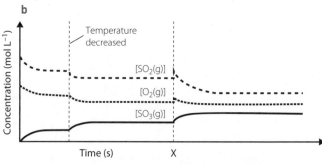

At time X, when the volume of the reaction vessel is decreased, the concentrations of all species simultaneously increase by the same factor. Then equilibrium shifts to the side with fewer gaseous particles, the product side, to decrease the pressure according to Le Chatelier's principle. The decrease in concentration of SO_2 and O_2 is in a $2:1$ ratio as observed in the mole coefficients in the balanced equation, while the increase in concentration of SO_3 is in the same ratio as SO_2.

3 X is a catalyst because it increases the rate of both the forward and reverse reactions. This is because the reaction takes an alternative pathway with a lower activation energy, which results in an increase in the percentage of successful collisions per unit time. The activation energies of both the forward and reverse reactions must be low enough so that enough particles will have sufficient energy for successful collisions; hence, the rate of both forward and reverse reactions increase.

4
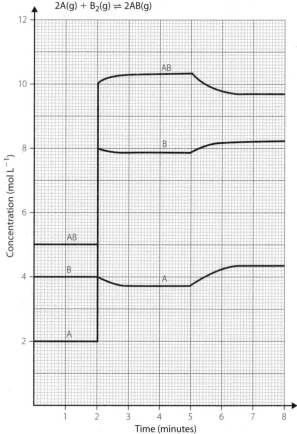

9780170449656

Since the system was at equilibrium between 0 and 2 minutes, the concentrations of A, B_2 and AB were constant at 2, 4 and $5 \, mol \, L^{-1}$ respectively. At 2 minutes, the volume of the system was halved; therefore, the concentrations of all three gases A, B_2 and AB doubled, i.e. to 4, 8 and $10 \, mol \, L^{-1}$ respectively. Equilibrium shifted to the right, the product side, increasing the concentration of AB while the concentrations of A and B_2 decreased. The ratio of the change was $2AB:2A:1B_2$. A new equilibrium was established at 3.0 minutes; therefore, the concentrations of all species remain constant. The temperature was increased at 5 minutes and equilibrium shifts to the left, the reactant side, in the endothermic direction. The concentration of AB decreases while concentrations of A and B_2 increase, again in the ratio $2AB:2A:1B$. A new equilibrium was established at 6.5 minutes; therefore, the concentrations of all species remained constant until 8 minutes.

Chapter 2: Factors that affect equilibrium

WS 2.1 PAGE 15

1 a False. At dynamic equilibrium the macroscopic properties are constant.

   b False. The concentrations of reactants and products are not the same, but they are constant.

   c True

   d False. When the reaction between nitrogen gas and hydrogen gas to produce ammonia reaches equilibrium, nitrogen gas, hydrogen gas and ammonia gas are all present in the reaction vessel.

   e True

2 a The equilibrium mixture will be brown in colour at room temperature. This is because it will be a mixture of the two gases that are colourless and dark brown in colour; thus, the resulting colour will be brown.

   b The equilibrium as written is endothermic. Heating the mixture causes the equilibrium to shift to the right,

~continue in right column ▲

the product side, which produces the dark brown gas. Hence, according to Le Chatelier's principle, if a system is at equilibrium and a change is made that upsets the equilibrium, then the system alters in such a way as to counteract the change and a new equilibrium is established. If the temperature is increased, equilibrium shifts to the side that leads to the absorption of the extra heat and a decrease in temperature, i.e. the endothermic direction.

   c The plunger in the syringe was pushed in to increase the pressure. This is because, if the pressure of a gaseous system is increased, equilibrium shifts to the side with fewer gaseous molecules, which is the left-hand side or reactant side with one gaseous molecule versus two gaseous molecules on the product side. Since more of the colourless N_2O_4 is produced, the brown colour becomes lighter.

3 a The equilibrium mixture will be blue-green in colour at room temperature. This is because it will be a mixture of the two coloured species that are blue and green in colour, so the resulting colour will be blue-green.

   b The blue-green colour will become more green. According to Le Chatelier's principle, if a system is at equilibrium and a change is made that upsets the equilibrium, then the system alters in such a way as to counteract the change and a new equilibrium is established. In this case, the addition of dilute hydrochloric acid would result in an increase in the concentration of Cl^-, resulting in a shift of the equilibrium in the direction that uses up the Cl^- to reduce its concentration, i.e. to the right-hand product side.

   c Add a compound such as silver nitrate solution that will precipitate Cl^- out of solution according to the reaction $Ag^+(aq) + Cl^-(aq) \rightarrow AgCl(s)$, thereby reducing the concentration of Cl^-. According to Le Chatelier's principle, the system will adjust to replace the Cl^- ions that precipitated, resulting in the equilibrium shifting to the left-hand side, the reactant side, which has $Cu(H_2O)_4^{2+}$, which is blue in colour.

d

Change	Observation	Effect of change	Explanation
Solid I_2 is added.	No visible change.	No change.	Since I_2 is a pure solid, adding more does not change its concentration; therefore, there is no change in equilibrium.
Gaseous I_2 is added.	Purple colour becomes lighter, while more black solid forms.	Reverse reaction is favoured.	At equilibrium, the rate of the forward reaction equals the rate of the reverse reaction. Increased concentration of gaseous I_2 results in more collisions and an increased rate of the reverse reaction. Adding more gas also means increased pressure, so equilibrium shifts to the left to the reactant side to reduce the increased pressure.
Volume of the reaction container is decreased.	Purple colour becomes lighter, while more black solid forms.	Reverse reaction is favoured.	A decrease in volume favours the side with fewer gaseous molecules because there is an increased concentration of gaseous I_2; hence, there are more collisions between gaseous I_2 which results in an increased rate of the reverse reaction. The shift to the reactant side would also result in a decrease in pressure.

Chapter 3: Calculating the equilibrium constant (K_{eq})

WS 3.1 PAGE 18

1 a i True **ii** False **iii** False
iv True **v** False

b ii $K_{eq} = \dfrac{[CO][H_2]^3}{[CH_4][H_2O]}$

iii $K > 1$ means the equilibrium lies to the right, as there is a higher concentration of products.

v Increasing the concentration of reactants will not affect the equilibrium constant. (Only a change to the temperature will change the equilibrium constant.)

2 a $K_{eq} = \dfrac{[Hb(O_2)_4]}{[Hb][O_2]^4}$

b i $Hb(aq) + 4CO(aq) \rightleftharpoons Hb(CO)_4(aq)$

ii $K_{eq} = \dfrac{[Hb(CO)_4]}{[Hb][CO]^4}$

c i $Hb(CO)_4(aq) + 4O_2(aq) \rightleftharpoons Hb(O_2)_4(aq) + 4CO(aq)$

ii $K_{eq} = \dfrac{[Hb(O_2)_4][CO]^4}{[Hb(CO)_4][O_2]^4}$

3 When it is in a gaseous state, as water vapour, or when water is acting as a reactant, not just a solvent.

4 a $H_2(g) + Cl_2(g) \rightleftharpoons 2HCl(g)$

$K_{eq} = \dfrac{[HCl]^2}{[H_2][Cl_2]}$

b Addition of heat would shift the reaction to the left as it would move to absorb the extra heat and thus lower the temperature of the system. This shift would result in higher concentrations of reactants and decrease the concentration of the products, therefore, the K_{eq} decreases.

c The equilibrium constant is the ratio of products to reactants for a specific temperature. When the temperature is changed, the system will adjust to minimise the change. This leads to a new ratio as equilibrium is re-established and so a new value for the equilibrium constant.

d $K_{eq} = \dfrac{[H_2][Cl_2]}{[HCl]^2}$

e If both the forward and reverse reactions occur at the same temperature, then the K_{eq} of the reverse reaction will be the reciprocal of K_{eq} of the forward reaction.

WS 3.2 PAGE 20

1 a, b, d, e

	2A(g)	⇌ 4B(g)	+ C(g)
Initial concentration	0.500	0.000	0.000
Change in concentration	(0.500 − 0.350) = −0.150	$\frac{4}{2}\times0.150$ = +0.3	$\frac{1}{2}\times0.150$ = +0.075
Equilibrium concentration	0.350	0.300	0.075

c $A = -x$; $B = +2x$; $C = \dfrac{1}{2}x$

f $K_{eq} = \dfrac{[B]^4[C]}{[A]^2}$

g $K_{eq} = \dfrac{[B]^4[C]}{[A]^2} = \dfrac{[0.300]^4[0.075]}{[0.350]^2} = 4.96 \times 10^{-3}$

2 a

	2NO(g) +	Cl$_2$(g)	⇌ 2NOCl(g)
Initial concentration	2.63	1.92	0.00
Change in concentration	−1.96	$\frac{1}{2}\times1.96 = -0.98$	+1.96
Equilibrium concentration	0.67	0.94	1.96

b $K_{eq} = \dfrac{[NOCl]^2}{[NO]^2[Cl_2]} = \dfrac{[1.96]^2}{[0.67]^2[0.94]} = 9.10$

3 a

	NH$_4$Cl(s) ⇌	NH$_3$(g) +	HCl(g)
Initial concentration		0.00	0.00
Change in concentration		+1.42	+1.42
Equilibrium concentration		1.42	1.42

b Solids are not included in the equilibrium expression and therefore do not need to be calculated.

c $K_{eq} = [NH_3][HCl] = [1.42][1.42] = 2.02$

4

	N$_2$(g) +	3H$_2$(g) ⇌	2NH$_3$(g)
Initial concentration	2.70	9.21	0.00
Change in concentration	$\frac{1}{3}\times7.86 = -2.62$	(9.21 − 1.35) = 7.86	$+\frac{2}{3}\times7.86 = 5.24$
Equilibrium concentration	0.08	1.35	5.24

$K_{eq} = \dfrac{[NH_3]^2}{[N_2][H_2]^3} = \dfrac{[5.31]^2}{[0.08][1.35]^3} = 143$

5

	4HCl(g) +	O$_2$(g) ⇌	2Cl$_2$(g) +	2H$_2$O(g)
Initial concentration	8.8	7.55	0.000	0.000
Change in concentration	−5.0	−1.25	(2 × 1.25) = 2.5	(2 × 1.25) = 2.5
Equilibrium concentration	3.8	6.3	2.5	2.5

$K_{eq} = \dfrac{[Cl_2]^2[H_2O]^2}{[HCl]^4[O_2]} = \dfrac{[2.5]^2[2.5]^2}{[3.8]^4[6.3]} = 0.030$

6

	NH$_4$HS(s) ⇌	NH$_3$(g) +	H$_2$S(g)
Initial concentration		0.00	0.00
Change in concentration		0.168	0.168
Equilibrium concentration		$c = \dfrac{n}{V} = \dfrac{0.418}{2.50} = 0.168$	0.168

$K_{eq} = [NH_3][H_2S] = [0.168][0.168] = 0.0282$

7

	$N_2(g)$	$+$	$2O_2$	$\rightleftharpoons$	$2NO_2(g)$
Initial concentration	0.036		$c=\dfrac{n}{V}=\dfrac{0.089}{2.0}$ $=0.0445$		0.00
Change in concentration	$\dfrac{0.0126}{2}=$ 0.0063		0.0126		0.0126
Equilibrium concentration	0.0297		0.0319		0.0126

$$K_{eq}=\frac{[NO_2]^2}{[N_2][O_2]^2}=\frac{[0.0126]^2}{[0.0297][0.0319]^2}=5.3$$

8

	$PCl_5(g)$	$\rightleftharpoons$	$PCl_3(g)$	$+$	$Cl_2(g)$
Initial concentration	$c=\dfrac{n}{V}=\dfrac{0.4}{10}$ $=0.04$		0.00		0.00
Change in concentration	0.025		0.025		0.025
Equilibrium concentration	0.015		0.025		$c=\dfrac{n}{V}=\dfrac{0.25}{10}$ $=0.025$

$$K_{eq}=\frac{[PCl_3][Cl_2]}{[PCl_5]}=\frac{[0.025][0.025]}{[0.015]}=0.042$$

WS 3.3 PAGE 23

1 a $K_{eq}=\dfrac{[CO][Br_2]}{[COBr_2]}=\dfrac{0.17\times0.17}{0.15}=0.19$

b $K_{eq}=\dfrac{[SO_2]^2[O_2]}{[SO_3]^2}=\dfrac{0.25^2\times0.86}{0.37^2}=0.39$

c $K_{eq}=\dfrac{1}{[SO_3][H_2O]}=\dfrac{1}{0.400\times0.480}=5.21$

Liquids and solids are excluded from equilibrium expressions as their concentrations remain constant.

d $[H_2O]=\dfrac{0.050}{2}=0.025$ mol L^{-1}

$[HCl]=\dfrac{0.750}{2}=0.375$ mol L^{-1}

$[POCl_3]=\dfrac{0.500}{2}=0.250$ mol L^{-1}

$K_{eq}=\dfrac{[HCl]^2[POCl_3]}{[H_2O]}=\dfrac{0.375^2\times0.250}{0.025}=1.406$

2 a $A_2+2B_2\rightleftharpoons2AB_2$

b 5 minutes

c $c(A_2)=\dfrac{n}{V}=\dfrac{2}{2.3}=0.87$ mol L^{-1}

$c(B_2)=\dfrac{n}{V}=\dfrac{1.2}{2.3}=0.52$ mol L^{-1}

$c(AB_2)=\dfrac{n}{V}=\dfrac{0.8}{2.3}=0.35$ mol L^{-1}

$K_{eq}=\dfrac{[AB_2]^2}{[A_2][B_2]^2}=\dfrac{0.35^2}{0.87\times0.52^2}=0.52$

d $c(A_2)=\dfrac{n}{V}=\dfrac{1.7}{2.3}=0.74$ mol L^{-1}

$c(B_2)=\dfrac{n}{V}=\dfrac{0.6}{2.3}=0.26$ mol L^{-1}

$c(AB_2)=\dfrac{n}{V}=\dfrac{1.4}{2.3}=0.61$ mol L^{-1}

$K_{eq}=\dfrac{[AB_2]^2}{[A_2][B_2]^2}=\dfrac{0.61^2}{0.74\times0.26^2}=7.44$

e As the K_{eq} values changed, increasing by 6.92, the temperature of the system must have been changed.

f A K_{eq} value less than 1 indicates that the equilibrium lies to the left; therefore, the yield of the product is low.

3 a $K_{eq}=\dfrac{[PCl_3][Cl_2]}{[PCl_5]}=\dfrac{1.3\times10^{-2}\times3.9\times10^{-3}}{4.2\times10^{-5}}=1.2$

b $1.2=\dfrac{1.0\times10^{-2}\times1.0\times10^{-2}}{x}$

$x=\dfrac{1.0\times10^{-2}\times1.0\times10^{-2}}{1.2}=8.3\times10^{-5}$ mol L^{-1}

4 $K_{eq}=\dfrac{[NH_3]^2}{[N_2][H_2]^3}$

$0.080=\dfrac{x^2}{0.600\times0.420^3}$

$x^2=0.080\times0.600\times0.420^3$

$=0.0036$

$\therefore x=\sqrt{0.0036}=0.060$ mol L^{-1}

5 a $K_{eq}=\dfrac{[H^+][F^-]}{[HF]}=\dfrac{0.0075\times0.0075}{0.085}=0.00066$

b $[F^-]=[H^+]=0.010$ mol L^{-1}

$0.00066=\dfrac{0.010\times0.010}{x}$

$x=\dfrac{0.010\times0.010}{0.00066}=0.15$ mol L^{-1}

Chapter 4: Solution equilibria

WS 4.1 PAGE 26

1 a $NaCl(s)\rightarrow Na^+(aq)+Cl^-(aq)$

b Sodium chloride is an ionic compound with ionic bonds between ions in an ionic lattice, and water is a polar molecule with hydrogen bonds between molecules. The partial negative charge on the oxygen atom in water attracts the sodium ion, while the partial positive charge on the hydrogen atoms attracts the chloride ion. The ionic bonds in the ionic lattice are broken as are the hydrogen bonds between water molecules. New ion–dipole bonds are formed between the water molecules and the dissociated sodium and chloride ions.

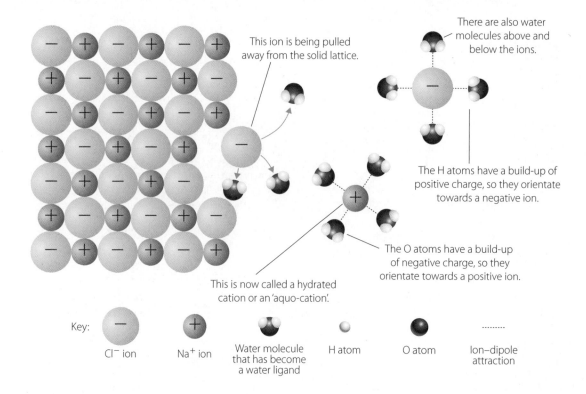

Key:

⊖	⊕		○	●	·········
Cl⁻ ion	Na⁺ ion	Water molecule that has become a water ligand	H atom	O atom	Ion–dipole attraction

This ion is being pulled away from the solid lattice.

There are also water molecules above and below the ions.

The H atoms have a build-up of positive charge, so they orientate towards a negative ion.

The O atoms have a build-up of negative charge, so they orientate towards a positive ion.

This is now called a hydrated cation or an 'aquo-cation'.

c Energy is needed to break the ionic bonds between sodium ions and chloride ions in the sodium chloride lattice; energy is also needed to break the hydrogen bonds between water molecules. Energy is released when ion–dipole bonds form between the water molecules and the ions; that is, as the charged ions become hydrated. The entropy of the system increases as the ordered sodium chloride lattice dissolves and the hydrated ions move randomly through the solution.

d i $\Delta H^{\theta} = \Delta H^{\theta}\left(Na^{+}(aq)\right) + \Delta H^{\theta}\left(Cl^{-}(aq)\right) - \Delta H^{\theta}\left(NaCl(s)\right)$

$= (-240) + (-167) - (-411)$

$= +4 \text{ kJ mol}^{-1}$

The dissolution of sodium chloride is endothermic. This means the energy required to break the ionic bonds in the NaCl lattice is greater than the energy released when ion–dipole bonds form between water and the ions.

ii $\Delta S^{\theta} = S^{\theta}\left(Na^{+}(aq)\right) + S^{\theta}\left(Cl^{-}(aq)\right) - S^{\theta}\left(NaCl(s)\right)$

$= 59 + 56.5 - 72.1$

$= +43.4 \text{ J K}^{-1} \text{ mol}^{-1}$

$= 0.0434 \text{ kJ K}^{-1} \text{ mol}^{-1}$

The entropy (randomness) of the system has increased, as shown by the positive value for ΔS^{θ}.

iii $\Delta G^{\theta} = \Delta H^{\theta} - T\Delta S^{\theta}$

$= 4 - (298 \times 0.0434)$

$= -8.93 \text{ kJ mol}^{-1}$

As ΔG^{θ} is negative, the reaction will proceed. spontaneously in the forward direction at 25°C.

2 a i 30 g of $KClO_3$ will dissolve in 100 g water, so 15 g will dissolve in 50 g water at 70°C.

ii 63 g of $CaCl_2$ will dissolve in 100 g water, so 94.5 g will dissolve in 150 g water at 10°C.

b i 19 g

ii At 50°C, 87 g will dissolve. 87 − 19 = 68 g
The extra amount that would dissolve at 50°C is 68 g.

c i The minimum temperature at which 40 g will dissolve, i.e. 62°C.

ii Above this temperature, all of the salt will have dissolved so the solution will be unsaturated.

d Generally, as temperature increases, more of the salt will dissolve. This is not true for $Ce_2(SO_4)_3$ where the solubility of this salt decreases with temperature.

3 a

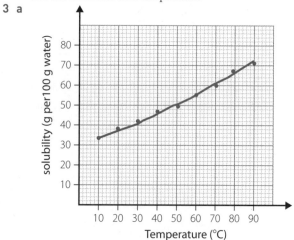

b $NH_4Cl(s) \rightarrow NH_4^{+}(aq) + Cl^{-}(aq)$

c i 43 g

ii $M(NH_4Cl) = 14.01 + 1.008 \times 4 + 35.45$
$= 53.492 \text{ g mol}^{-1}$

$n = \dfrac{43}{53.492} = 0.804 \text{ mol } V = 100 \text{ mL} = 0.100 \text{ L}$

$c = \dfrac{n}{V} = \dfrac{0.804}{0.100} = 8.04 \text{ mol L}^{-1} = 8.0 \text{ mol L}^{-1}$

d From the graph, 58 g of NH_4Cl will dissolve in 100 g water at 65°C.

Mass of water needed to dissolve 36 g $= \dfrac{36}{58} \times 100 = 62 \text{ g}$

9780170449656

1 a $Ag^+(aq) + Cl^-(aq) \rightarrow AgCl(s)$

b

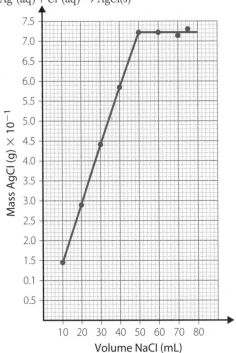

c The graph has two straight lines, one running from the 10 mL to the point 50 mL, 0.720 g and the other parallel to the x axis through 0.720 g.

d At 50 mL NaCl, all of the Ag^+ ion has been precipitated out, so using larger volumes of NaCl does not produce any extra AgCl precipitate.

e $n(AgCl) = \dfrac{0.720}{107.9 + 35.45} = 5.02 \times 10^{-3} \text{ mol}^{-1}$

$n(AgNO_3) = 5.02 \times 10^{-3} \text{ mol}^{-1}$

f $n(AgNO_3) = 5.02 \times 10^{-3} \text{ mol}^{-1}$ $V = 50 \text{ mL} = 0.050 \text{ L}$

$c(AgNO_3) = \dfrac{5.02 \times 10^{-3} \text{ mol}^{-1}}{0.050} = 0.100 \text{ mol L}^{-1}$

2 a Precipitate; $Cu^{2+}(aq) + 2OH^-(aq) \rightarrow Cu(OH)_2(s)$

b Precipitate; $2Al^{3+}(aq) + 3CO_3^{2-}(aq) \rightarrow Al_2(CO_3)_3(s)$

c No precipitate

3 a A is Cl^- as it has only formed a precipitate with Ag^+ and this is consistent with the solubility rules, $Ag^+(aq) + Cl^-(aq) \rightarrow AgCl(s)$.

B is NO_3^-, the solubility rules indicate NO_3^- does not form any precipitates and this is consistent with the results for B.

C is SO_4^{2-}; according to the solubility rules SO_4^{2-} forms precipitates with Ba^{2+} and Ca^{2+} and Ag^+ but not with Al^{3+} and this is consistent with the results.

D is O^{2-}; according to the solubility rules O^{2-} does not form precipitates with Ba^{2+} and Ca^{2+} but does form precipitates with Ag^+ and Al^{3+} and this is consistent with the results for D.

b

Cation added	A	B	C	D	CH₃COO⁻	CO₃²⁻
Ba^{2+}	NP	NP	ppt	NP	NP	ppt
Ag^+	ppt	NP	ppt	ppt	ppt	ppt
Ca^{2+}	NP	NP	ppt	NP	NP	ppt
Al^{3+}	NP	NP	NP	ppt	NP	ppt

4

```
                    ┌──────────┐
                    │  Sample  │
                    └──────────┘
                         │ Add KI & filter
          Filtrate ◄─────┴─────► Residue
                                    │ Ppt
                          Pb²⁺(aq) + 2I⁻(aq) → PbI₂(s)
   Add Na₂SO₄ & filter
          Filtrate ◄─────┴─────► Residue
                                    │ Ppt
                          Ca²⁺(aq) + SO₄²⁻(aq) → CaSO₄(s)
   Add NaOH & filter
          Filtrate ◄─────┴─────► Residue
                                    │ Ppt
   Filtrate               Mg²⁺(aq) + 2OH⁻(aq) → Mg(OH)₂(s)
```

5 a $Pb^{2+}(aq) + 2I^-(aq) \rightarrow PbI_2(s)$

b $n(PbI_2) = \dfrac{m}{M} = \dfrac{0.0154}{(207.2 + 2 \times 126.9)} = 3.341 \times 10^{-5} \text{ mol}$

$n(Pb^{2+}) = n(PbI_2) = 3.341 \times 10^{-5} \text{ mol}$

c $m(Pb^{2+})$ in 100 mL $= 3.341 \times 10^{-5} \times 207.2 = 6.92 \times 10^{-3} \text{ g}$

$m(Pb^{2+})$ in 1 L $= 6.92 \times 10^{-2} \text{ g} = 69.2 \text{ mg}$

The EPA limit is 0.01 mg L⁻¹, so the amount of lead in the water is far above the safe limit.

d They assumed no other cations in the sample would precipitate with KI. If Ag^+ was in the sample, it would also precipitate. This means the sample would not be pure PbI_2 so the amount of lead present would be less than that calculated.

They assumed the lead was evenly mixed in the river. If this is not so, there may be more or less lead present in the river. They also assumed that all the lead ions precipitated out when the KI was added. If this was not correct (that is, the KI was not in excess), then there would be more lead present in the river than the analysis determined.

1 a To determine the K_{sp} of calcium hydroxide

b i As calcium hydroxide is a slightly soluble salt only a small amount will dissolve. It is the dissolved ions which are being analysed so the actual amount of solid added does not need to be known. It just needs to be more than the molar solubility of the compound.

ii Solubility $= 0.0108 \text{ mol L}^{-1}$

Molar mass $Ca(OH)_2 = 40.08 + 2 \times (16.00 + 1.008)$
$= 74.096 \text{ g mol}^{-1}$

Solubility in g L⁻¹ $= 74.096 \times 0.0108 = 0.8002 \text{ g L}^{-1}$

Yes, they added enough $Ca(OH)_2$ because they only made 100 mL, so they only needed to add 0.08 g.

iii $Ca(OH)_2(s) \rightleftharpoons Ca^{2+}(aq) + 2OH^-(aq)$

iv According to the equilibrium equation, the $Ca(OH)_2(s)$ is in equilibrium with $OH^-(aq)$. Adding HCl would remove $OH^-(aq)$ according to the reaction $H^+(aq) + OH^-(aq) \rightarrow H_2O(l)$. According to Le Chatelier's principle, more solid calcium hydroxide would dissolve in order to increase the concentration of $OH^-(aq)$, to partially replace what had reacted. As the analysis is to find out how much $OH^-(aq)$ is present in the saturated solution, the solid needs to be removed so no additional dissolving can occur. If it was not removed, the concentration of $OH^-(aq)$ would be much greater than in the saturated solution.

c **i** The average is calculated by adding the volumes then dividing the total by the number of volumes added. In this case, the volume for flask 1 is not included as it is an anomalous result when considering the values for flasks 2, 3 and 4.

$$\text{Average} = \frac{2.30 + 2.20 + 2.40}{3} = 2.30\,\text{mL}$$

ii $n(\text{HCl}) = c \times V \quad c = 0.100\,\text{mol L}^{-1}, V = 2.3\,\text{mL} = 0.00230\,\text{L}$

$n = 0.100 \times 0.00230 = 0.000230\,\text{mol} = 2.30 \times 10^{-4}\,\text{mol}$

iii $n(\text{HCl}) = 2.30 \times 10^{-4}\,\text{mol}$

$n(\text{HCl}) = n(\text{H}^+) = n(\text{OH}^-) = 2.30 \times 10^{-4}\,\text{mol}$

$V(\text{OH}^-) = 10\,\text{mL} = 0.01\,\text{L}^{-1}$

$c(\text{OH}^-) = \frac{n}{V} = \frac{2.30 \times 10^4}{0.01} = 2.30 \times 10^{-2} = 0.023\,\text{mol L}^{-1}$

d $\text{Ca(OH)}_2(\text{s}) \rightleftharpoons \text{Ca}^{2+}(\text{aq}) + 2\text{OH}^-(\text{aq})$

$n(\text{Ca}^{2+}) = \frac{1}{2}n\text{OH}^- = \frac{2.30 \times 10^{-4}}{2} = 1.15 \times 10^{-4}\,\text{mol}$

$c(\text{Ca}^{2+}) = \frac{n}{V} = \frac{1.15 \times 10^{-4}}{0.01} = 1.15 \times 10^{-2} = 0.0115\,\text{mol}$

e $K_{sp} = [\text{Ca}^{2+}] \times [\text{OH}^-]^2 = (0.0115) \times (0.023)^2 = 6.08 \times 10^{-6}$

f The theoretical K_{sp} value for $\text{Ca(OH)}_2 = 5.02 \times 10^{-6}$. The experimental value is slightly larger than the theoretical value; however, it is still in the same order of magnitude. Possible reasons for the difference would be experimental errors such as:

▶ a pipette should have been used to measure an accurate volume of 10 mL Ca(OH)_2 that was placed in the flasks as using a measuring cylinder is less accurate

▶ the neutralisation point was misjudged

▶ K_{sp} value is for 25°C so temperature conditions should have been considered.

WS 4.4 PAGE 37

1 Yes, it is correct to say that a saturated solution of an ionic solid containing excess solid is in a state of dynamic equilibrium. The macroscopic properties of the solution are constant, the concentration of the ions in solution is constant so the ions are leaving and returning to the solid at equal rates. However, the process of dissolving an ionic solid in water is usually regarded as a physical process rather than a chemical one because the solid can be recovered through evaporation so there is no chemical change. Precipitation reactions produce a new product so are seen as chemical reactions. The resultant insoluble or sparingly soluble salt also reaches dynamic equilibrium. Solubility products can be calculated for any salt; however, very soluble salts, for example sodium chloride, are difficult to precipitate from solution so solubility products are rarely used.

2 **a** Barium hydroxide > calcium hydroxide > magnesium hydroxide

b The hydroxides become more soluble as you go down the group.

c The solubility of the group 2 carbonates decreases as you go down the group with the order of solubility from most to least being: magnesium carbonate > calcium carbonate > barium carbonate.

d No generalisation can be made about the solubility of group 2 salts. The solubility of the hydroxide salts increases as you go down the group, while the solubility of the carbonate salts decreases as you go down the group. The solubility of the phosphate salts goes up then down, so there is no overall pattern to the solubility of different salts for this group.

3 **a** $\text{AgI}(\text{s}) \rightleftharpoons \text{Ag}^+(\text{aq}) + \text{I}^-(\text{aq})$

Solubility AgI = 2.11×10^{-7} g/100 g = 2.11×10^{-6} g L^{-1}

Molar mass AgI = $107.9 + 126.9 = 234.8$ g mol^{-1}

$n = \frac{2.11 \times 10}{234.8} = 8.99 \times 10$

Solubility of AgI = 8.99×10^{-9} mol L^{-1}

$[\text{Ag}^+] = [\text{I}^-] = 8.99 \times 10^{-9}$ mol L^{-1}

$K_{sp} = [\text{Ag}^+] \times [\text{I}^-] = (8.99 \times 10^{-9})^2 = 8.08 \times 10^{-17}$

b Volume is doubled, so concentration of each solution is halved.

$[\text{AgNO}_3] = \frac{1.46 \times 10^{-9}}{2} = 7.3 \times 10^{-10} = [\text{Ag}^+$

$[\text{MgI}_2] = \frac{2.02 \times 10^{-6}}{2} = 1.01 \times 10^{-6}$ mol L^{-1}

$[\text{I}^-] = 2 \times [\text{MgI}_2] = 2.02 \times 10^{-6}$ mol L^{-1}

$Q_{sp} = [\text{Ag}^+] \times [\text{I}^-] = (7.3 \times 10^{-10}) \times (2.02 \times 10^{-6})$
$= 1.47 \times 10^{-15}$ mol L^{-1}

$Q_{sp} > K_{sp}$ so a precipitate will form.

4 **a** $\text{CaCO}_3(\text{s}) \rightleftharpoons \text{Ca}^{2+}(\text{aq}) + \text{CO}_3{}^{2-}(\text{aq})$

Let $[\text{Ca}^{2+}] = s\,\text{mol L}^{-1}$ and $\text{CO}_3{}^{2-} = s\,\text{mol L}^{-1}$

$K_{sp} = [\text{Ca}^{2+}] \times \text{CO}_3{}^{2-} = 3.36 \times 10^{-9}$

$s \times s = 3.36 \times 10^{-9}$

$s^2 = 3.36 \times 10^{-9}$

$s = \sqrt{3.36 \times 10^{-9}} = 5.80 \times 10^{-5}$

$[\text{Ca}^{2+}] = 5.80 \times 10^{-5}\,\text{mol L}^{-1}$ and $\text{CO}_3{}^{2-} = 5.80 \times 10^{-5}\,\text{mol L}^{-1}$

b **i** $\text{CaCO}_3(\text{s}) \rightleftharpoons \text{Ca}^{2+}(\text{aq}) + \text{CO}_3{}^{2-}(\text{aq}$ is exothermic, so as the temperature decreases the forward reaction will proceed at a greater rate to compensate for the change, and according to Le Chatelier's principle, the exothermic reaction will be favoured to produce heat, and thus the CaCO_3 will dissolve.

ii A higher K_{sp} at lower depths means the $\text{CaCO}_3(\text{s})$ is more soluble because a higher K_{sp} means that more products (ions) are present and less solid (CaCO_3), so this will also contribute to the shells dissolving.

5 **a** $\text{PbCl}_2(\text{s}) \rightleftharpoons \text{Pb}^{2+}(\text{aq}) + 2\text{Cl}^-(\text{aq})$

$\text{AgCl}(\text{s}) \rightleftharpoons \text{Ag}^+(\text{aq}) + \text{Cl}^-(\text{aq})$

Let $[\text{Cl}^-] = s$,

$[\text{Pb}^{2+}] = 0.125\,\text{mol L}^{-1}$

$[\text{Ag}^+] = 0.000\,750\,\text{mol L}^{-1}$

For PbCl_2

$K_{sp} = [\text{Pb}^{2+}] \times [\text{Cl}^-]^2 = 1.70 \times 10^{-5}$

$(0.125) \times (s^2) = 1.70 \times 10^{-5}$

$s^2 = \frac{1.70 \times 10^{-5}}{0.125} = 1.36 \times 10^{-4}$

$s = \sqrt{1.36 \times 10^{-4}} = 0.0117\,\text{mol L}^{-1}$

For AgCl

$K_{sp} = [\text{Ag}^+] \times [\text{Cl}^-] = 1.77 \times 10^{-10}$

$(7.50 \times 10^{-4}) \times (s) = 1.77 \times 10^{-10}$

$s = \frac{1.77 \times 10^{-10}}{7.50 \times 10^{-4}} = 2.36 \times 10^{-7}\,\text{mol L}^{-1}$

For PbCl_2 to form, $[\text{Cl}^-]$ must be at least $0.0117\,\text{mol L}^{-1}$.

For AgCl to form, $[\text{Cl}^-]$ must be at least $2.36 \times 10^{-7}\,\text{mol L}^{-1}$.

AgCl will precipitate first as it needs a much lower concentration of chloride ions.

b As AgCl will precipitate first, enough Cl$^-$ ions need to be added to precipitate out all the Ag$^+$ ions but there needs

9780170449656

to be less than $0.0117\,mol\,L^{-1}$ of Cl^- ions so Pb^{2+} does not precipitate.

In 100 mL of solution:

$$n_{Ag^+} = cV = 7.50 \times 10^{-4} \times 0.100 = 7.50 \times 10^{-5}\,mol$$

Therefore, to precipitate all the Ag^+, the number of moles Cl^- added. $n(Cl^-) = 7.50 \times 10^{-5}\,mol$, The concentration of the solution will be dependent on the volume used.

Suppose 10 mL HCl is used.

$$[HCl] = [Cl^-] = \frac{7.50 \times 10^{-5}}{0.01} = 0.0075\,mol\,L^{-1}$$

This concentration is less than the concentration needed for Pb^{2+} to precipitate.

Method

Add 10 mL $0.0075\,mol\,L^{-1}$ HCl to 100 mL of solution, allow AgCl to precipitate, then filter to remove precipitate, leaving filtrate, which contains Pb^{2+} ions.

6 a $Ba_3(PO_4)_2(s) \rightleftharpoons 3Ba^{2+}(aq) + 2PO_4^{3-}(aq)$

$[Ba^{2+}] = 3s\,mol\,L^{-1}$; $[PO_4^{3-}] = 2s\,mol\,L^{-1}$

$K_{sp} = [Ba^{2+}]^3 \times [PO_4^{3-}]^2 = 1.30 \times 10^{-29}$

$(3s)^3 \times (2s)^2 = 1.30 \times 10^{-29}$

$108s^5 = 1.30 \times 10^{-29}$

$$s^5 = \frac{1.30 \times 10^{-29}}{108} = 1.20 \times 10^{-31}$$

$$s = \sqrt[5]{1.20 \times 10^{-31}} = 6.54 \times 10^{-7}$$

$[Ba^{2+}] = 3s = 1.96 \times 10^{-6}\,mol\,L^{-1}$;

$[PO_4^{3-}] = 2s = 1.31 \times 10^{-6}\,mol\,L^{-1}$

b i Add a soluble compound containing phosphate ions, e.g. Na_3PO_4. This would increase the concentration of phosphate ions and drive the reverse reaction to decrease the concentration of phosphate ions according to Le Chatelier's principle. This would cause more $Ba_3(PO_4)_2(s)$ to form, thus removing barium ions from solution.

ii Add a soluble compound containing phosphate ions, e.g. Na_3PO_4. This would increase the concentration of phosphate ions. Although some would be used in producing more $Ba_3(PO_4)_2(s)$, the final concentration of phosphate ions would still be greater as, according to Le Chatelier's principle, the imposed change can only be partially reversed.

OR Add a compound containing an anion that will precipitate Ba^{2+}. The K_{sp} of this compound must be smaller than the K_{sp} of barium phosphate, i.e. the new barium compound must be less soluble than barium phosphate. This would drive the equilibrium to the right to replace barium ions removed from solution.

7 a $Na_2SO_4(s) \rightarrow 2Na^+(aq) + SO_4^{2-}(aq)$

$[Na_2SO_4] = [SO_4] = 0.200\,mol\,L^{-1}$

$Ag_2SO_4(s) \rightleftharpoons 2Ag^+(aq) + SO_4^{2-}(aq)$

$[Ag^+] = 2s$

$[SO_4^{2-}] = s + 0.200$ (due to dissociation of Na_2SO_4 and Ag_2SO_4)

Since $s \ll 0.200$, assume $[SO_4^{2-}] = 0.200$

$K_{sp} = [Ag^+]^2 \times [SO_4^{2-}] = 1.20 \times 10^{-5}$

$(2s)^2 \times (0.200) = 1.20 \times 10^{-5}$

$0.800 \times s^2 = 1.20 \times 10^{-5}$

$$s = \sqrt{\frac{1.20 \times 10^{-5}}{0.800}} = 3.87 \times 10^{-3}\,mol\,L^{-1}$$

$[Ag^+] = 2s = 7.74 \times 10^{-3}\,mol\,L^{-1}$

$n(Ag^+) = cV = 7.74 \times 10^3 \times 0.45 = 3.483 \times 10^{-3}\,mol$

$$n(Ag_2SO_4) = \frac{n(Ag^+)}{2} = \frac{3.483 \times 10^{-3}}{2} = 1.7415 \times 10^{-3}\,mol$$

$m(Ag_2SO_4) = n \times M = 1.7415 \times 10^{-3} \times 311.87 = 0.54312\,g = 0.543\,g$

b One assumption is that sodium sulfate completely dissociates. This is justified because sodium sulfate is classified as a soluble salt.

A second assumption is that $s \ll 0.200$. To check the validity of this, calculate Q_{sp} using the concentrations $[Ag^+] = 7.74 \times 10^{-3}\,mol\,L^{-1}$ and $[SO_4^{2-}] = 0.200\,mol\,L^{-1}$.

$Q_{sp} = [Ag^+]^2 \times [SO_4^{2-}] = (7.74 \times 10^{-3})^2 \times (0.200) = 1.1982 \times 10^{-5}$
$= 1.2 \times 10^{-5}$

As $Q_{sp} = K_{sp}$, this assumption is also valid.

MODULE FIVE: CHECKING UNDERSTANDING PAGE 42

1 C   **2** B   **3** C   **4** A   **5** D
6 B   **7** C   **8** C   **9** D   **10** C

11 Any three of the following:

▸ Decrease the volume, which would increase the pressure. This would cause the system to increase pressure by shifting to the side of more molecules. According to Le Chatelier's principle, the forward reaction would be favoured to increase the number of gas molecules, so the amount of NO would increase.

▸ Increase the concentration of $NH_3(g)$, which, according to Le Chatelier's principle, would drive the forward reaction to decrease the increased $NH_3(g)$ concentration, producing more NO.

▸ Increase the concentration of $O_2(g)$, which, according to Le Chatelier's principle, would drive the forward reaction to decrease the increased $O_2(g)$ concentration, producing more NO.

▸ Decrease the temperature, which, according to Le Chatelier's principle, would drive the reaction in the exothermic (forward) direction in order to reduce the temperature.

12 The vapour pressure would rise until it reached 5.85 kPa and the level of liquid ethanol would decrease. The rate of evaporation would initially be greater than the rate of condensation but eventually a new equilibrium would be established.

13 a The system first reached equilibrium at 5 minutes, as can be seen by the constant concentration of all species present.

b More $Cl_2(g)$ was added to the system at the 10 minutes mark, as can be seen by the increase in concentration. This caused the forward reaction to be favoured according to Le Chatelier's principle to decrease the concentration of the $Cl_2(g)$, thus also decreasing the concentration of PCl_3 and increasing the concentration of PCl_5. Equilibrium was then re-established at approximately 14 minutes.

c

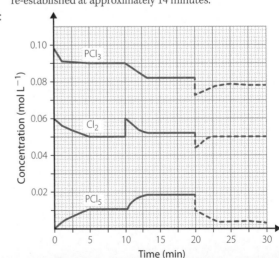

Increasing the volume would initially decrease the concentration of all the species by the same amount; for example, doubling the volume would halve the concentration. Increasing the volume would also cause a decrease in pressure. The rate of the reverse reaction would increase to partially increase the pressure as there are more gaseous molecules on the left and, by Le Chatelier's principle, the system will shift to increase pressure. Therefore, the concentration of Cl_2 and PCl_3 would increase, while the concentration of PCl_5 would decrease until equilibrium was re-established.

14 The higher partial pressure (hence concentration) of oxygen in the lungs than in the arterial blood causes oxygen to diffuse into the blood, causing reaction 2 to shift to the right, producing HbO_2 and H^+.

At the cell, CO_2 diffuses out due to differences in partial pressure, causing an increase in its concentration in the blood and causing reaction 1 to shift to the right, producing HCO_3^- (aq) and H^+(aq). The increase in H^+ at the cell causes reaction 2 to shift to the left, releasing O_2 into the blood, which then diffuses into the cell due to lower partial pressure.

15 a $K_{eq} = \dfrac{[CH_3CH_2OH]}{[CH_2=CH_2][H_2O]}$

b $[CH_2CH_2OH] = \dfrac{0.030}{5} = 0.006 \text{ mol L}^{-1}$

$[CH_2=CH_2] = \dfrac{0.50}{5} = 0.10 \text{ mol L}^{-1}$

$[H_2O] = \dfrac{1.0}{5} = 0.20 \text{ mol L}^{-1}$

$K_{eq} = \dfrac{0.006}{0.10 \times 0.20} = 0.30$

16 $CO(g) + Cl_2(g) \rightleftharpoons COCl_2(g)$

$K_{eq} = \dfrac{COCl_2}{[CO][Cl_2]}$

In the first vessel ($V = 1.0$ L),
$[COCl_2] = 1.5 \text{ mol L}^{-1}$; $[CO] = 0.15 \text{ mol L}^{-1}$; $[Cl_2] = 1.0 \text{ mol L}^{-1}$

$K_{eq} = \dfrac{1.5}{0.15 \times 1.0} = 10$

In the second vessel ($V = 1.0$ L),
$[COCl_2] = c \text{ mol L}^{-1}$; $[CO] = 0.10 \text{ mol L}^{-1}$; $[Cl_2] = 0.20 \text{ mol L}^{-1}$

$K_{eq} = 10 = \dfrac{c}{0.10 \times 0.2} = \dfrac{c}{0.02}$

~continue in right column ▲

$c = 10 \times 0.02 = 0.20$ $[COCl_2] = 0.20 \text{ mol L}^{-1}$
$MM(COCl_2) = 12.01 + 16.00 + 2 \times 35.45 = 98.91 \text{ g mol}^{-1}$
$m = n \times MM = 0.20 \times 98.91 = 89 \text{ g}$

17 a $Ba(OH)_2(s) \rightleftharpoons Ba^{2+}(aq) + 2OH^-(aq)$

b $K_{sp} = [Ba^{2+}] \times [OH^-]^2$
$[Ba^{2+}] = 0.108 \text{ mol L}^{-1}$; $[OH^-] = 2 \times 0.108 = 0.216 \text{ mol L}^{-1}$
$K_{sp} = (0.18) \times (0.216)^2 = 5.04 \times 10^{-3}$

c $[OH^-] = 0.225 \text{ mol L}^{-1}$
$[Ba^{2+}] = s \text{ mol L}^{-1}$
$K_{sp} = [Ba^{2+}] \times [OH^-]^2$
$5.04 \times 10^{-3} = s \times (0.225)^2$
$s = \dfrac{5.04 \times 10^{-3}}{(0.225)^2} = 0.0996$
$[Ba^{2+}] = 0.0996 \text{ mol L}^{-1}$

d $[Ba^{2+}] = \dfrac{0.210 \times 50}{100} = 0.105 \text{ mol L}^{-1}$

$[OH^-] = \dfrac{0.300 \times 50}{100} = 0.150 \text{ mol L}^{-1}$
$Q = 0.105 \times (0.150)^2 = 2.36 \times 10^{-3}$
$Q < K_{sp}$, so no precipitate will form.

MODULE SIX: ACID–BASE REACTIONS

REVIEWING PRIOR KNOWLEDGE PAGE 48

1 acid + base → salt + water

2 a $HCl(aq) + NaOH(aq) \rightarrow NaCl(aq) + H_2O(l)$
b $H_2CO_3(aq) + 2LiOH(aq) \rightarrow Li_2CO_3(aq) + 2H_2O(l)$
c $3Ca(OH)_2(aq) + 2H_3PO_4(aq) \rightarrow Ca_3(PO_4)_2(aq) + 6H_2O(l)$
d $2HBr(aq) + Ba(OH)_2(aq) \rightarrow BaBr_2(aq) + 2H_2O(l)$
e $Zn(OH)_2(aq) + 2HNO_3(aq) \rightarrow Zn(NO_3)_2(aq) + 2H_2O(l)$
f $Al(OH)_3(aq) + 3CH_3COOH(aq) \rightarrow Al(CH_3COO)_3(aq) + 3H_2O(l)$

3

Property	Acid	Base
Taste	Sour	Bitter
Texture (feel)	Sticky	Soapy
Change in litmus	Blue to red	Red to blue
Possible ions involved	H^+	OH^-

4

Subject	Formula to use	Calculation
Determine the number of moles in a 2.4 g sample of barium hydroxide.	$n = \dfrac{m}{MM}$	$n = \dfrac{2.4}{(137.3 + (16.00 + 1.008) \times 2)}$ $= \dfrac{2.4}{171.316}$ $= 0.014 \text{ mol}$
Calculate the mass of sodium hydroxide required to make a solution containing 0.135 moles.	$m = n \times MM$	$m = 0.135 \times (22.99 + 16.00 + 1.008)$ $= 0.135 \times 39.998$ $= 5.4 \text{ g}$
Determine the number of moles of hydrochloric acid in 22 mL of a 0.240 mol L^{-1} solution.	$n = c \times V$	$n = 0.240 \times 0.022$ $= 0.0053 \text{ moles}$
Calculate the concentration of a sodium chloride solution containing 0.125 moles sodium chloride in 500.0 mL of solution.	$c = \dfrac{n}{V}$	$c = \dfrac{0.125}{0.5000}$ $= 0.250 \text{ mol L}^{-1}$

Subject	Formula to use	Calculation
Calculate the mass of sodium hydroxide required to make a 0.51 mol L^{-1} solution in a 250 mL volumetric flask.	$n = c \times V$ $m = n \times MM$	$n = 0.51 \times 0.250$ $= 0.1275$ moles $m = 0.1275 \times (22.99 + 16.00 + 1.008)$ $= 0.1275 \times 39.998$ $= 5.1 \text{ g}$
What volume of 6 mol L^{-1} HCl is required in the preparation of 100 mL of 1 mol L^{-1} HCl?	$c_1V_1 = c_2V_2$	$V_2 = \dfrac{1 \times 0.1}{6}$ $= 0.017 \text{ L}$
You are supplied with 250 mL of a 0.4 mol L^{-1} solution. You add 250 mL of water to the solution. What is the concentration of the resulting solution?	$c_1V_1 = c_2V_2$	$c_2 = \dfrac{0.4 \times 0.250}{0.500}$ $= 0.2 \text{ mol L}^{-1}$

5

Calculate the number of moles of each reactant.	2
Write a balanced chemical equation.	1
Subtract the n of excess from the n of limiting.	4
Use the molar ratio to determine the limiting factor.	3

6 a $HCl(aq) + NaOH(aq) \rightarrow NaCl(aq) + H_2O(aq)$
$n(HCl) = 0.17 \times 0.025 = 0.00425$ moles
$n(NaOH) = 0.09 \times 0.05 = 0.0045$ moles
Ratio HCl : NaOH = 1 : 1; therefore, HCl is limiting.
Excess(NaOH) = 0.0045 − 0.00425 = 0.00025 moles

b $H_2SO_4(aq) + 2KOH(aq) \rightarrow K_2SO_4(aq) + 2H_2O(aq)$
$n(H_2SO_4) = 0.2 \times 0.0172 = 0.00344$ moles
$n(KOH) = 0.19 \times 0.032 = 0.00608$ moles
Ratio H_2SO_4 : KOH = 1 : 2; therefore, KOH is limiting as
$n(H_2SO_4)$ required is $\dfrac{0.00608}{2} = 0.00304$ moles.
Excess(H_2SO_4) = 0.00344 − 0.00304 = 0.0004 moles

7 In endothermic reactions, the energy within the bonds of the reactants is less than the energy within the products. The extra energy required to form the products is taken from the surrounding environment. We see this as a decrease in ambient temperature.

In exothermic reactions, the energy within the bonds of the reactants is higher than the energy within the products. The extra energy released from the breaking of the reactant bonds is lost to the environment. We see this as an increase in ambient temperature.

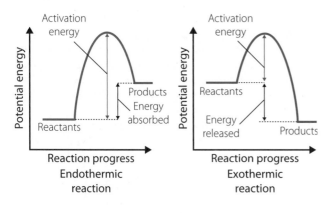

8 a

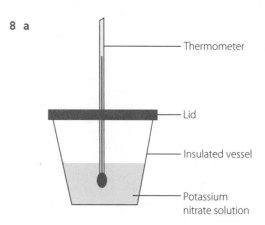

b

Volume of water (g)	100
$m(KNO_3)_3(g)$	4.80
Initial temperature (°C)	20.1
Final temperature (°C)	16.7
ΔT (°C)	−3.4

c $q = mC\Delta T = 0.1 \times 4.18 \times 10^3 \times 3.4 = 1421.2 \text{ J}$ (temperature drops so heat is absorbed and q is positive)

$$n(KNO_3) = \dfrac{m}{MM}$$
$$= \dfrac{4.8}{39.10 + 14.01 + 3 \times 16.00}$$
$$= 4.8 \times 101.11$$
$$= 0.04747 \text{ moles}$$

$\Delta H = \dfrac{q}{n} = \dfrac{1421.2}{0.04747} = +29\,936.99$ or $+29.9 \text{ kJ mol}^{-1}$

d If the calorimeter is not sufficiently insulated, energy from the surroundings will be absorbed as well as energy from the water. Therefore, the fall of the recorded water temperature will not be as large as expected.

9 a As per Le Chatelier's principle, the concentration of sulfuric acid will decrease as the reaction drives in the forward direction to increase the concentration of lead hydrogen sulfate to partially replace what has been removed.

b Solid lead sulfate is not included in the equilibrium expression. Adding more lead sulfate will result in more undissolved lead sulfate accumulating at the bottom of the beaker.

10 a $K_a = \dfrac{[H^+][C_6H_7O_6^-]}{[HC_6H_7O_6]}$

b Ascorbic acid ionises to a small extent. Since the equilibrium constant is so small, the equilibrium lies far to the left. That means lots of reactant remains in the beaker and very little product is formed.

c $8.0 \times 10^{-5} = \dfrac{[H^+][C_6H_7O_6^-]}{[HC_6H_7O_6]} = \dfrac{x \times x}{0.75}$

$x^2 = 0.75 \times 8.0 \times 10^{-5}$

$x = \sqrt{0.00006} = 7.5 \times 10^{-3}$

$[H^+] = 7.75 \times 10^{-3} \ mol \ L^{-1}$

Chapter 5: Properties of acids and bases

WS 5.1 PAGE 52

1

Indicator	Colour in pH < 7	pH range of colour change	Colour in pH > 7
Bromothymol blue	Yellow	6.0–7.6	Blue
Phenolphthalein	Colourless	8.3–10.0	Pink
Methyl orange	Red	3.1–4.4	Yellow
Litmus red	Red	4.5–8.3	Blue
Litmus blue	Red	4.5–8.3	Blue

2 a Beetroot turned yellow in the drain cleaner and provided a small change in colour for the baking soda. In the presence of lemon juice and vinegar, the beetroot did not change colour. This indicates that beetroot is useful to detect the presence of a strong base. However, due to the lack of distinct colour change in a weak base or in the presence of an acid, it would not be useful as an acid–base indicator.

  b 1 Collect red cabbage leaves and cut them into small pieces.

    2 Place the cabbage leaves in a beaker with water and boil until the water becomes coloured.

    3 Cool mixture.

    4 Decant the liquid into a container, leaving the solids behind.

  c The natural indicator should be added to samples of known acidic, neutral and basic solutions. The indicator is suitable if there is a discernible colour change across all solutions.

3 a Bromothymol blue would be the best indicator as it changes from yellow to blue in a narrow pH range of 6–7.5. Methyl orange is unsuitable as it would remain yellow across a large pH range, 4.5–13. Phenolphthalein is also unsuitable, remaining colourless until the pH is above the test range.

  b Barium sulfate is a neutral insoluble white powder. It absorbs the soil moisture and will show the indicator colour when it reacts with the moisture.

  c

pH	Colour of methyl orange	Colour of bromothymol blue
5.5	Yellow	Yellow
7	Yellow	Green
11.5	Yellow	Blue

  d The student's claim is invalid. Although the solution may have a pH of 5.5, due to the solution turning yellow in both indicators, the pH could range anywhere from 4.5 to 6.

4 Any three of:

  ▶ An indicator may change colour at a different pH when temperatures exceed 25°C.

  ▶ Indicator colour can be distorted if a solution is not colourless.

  ▶ Indicators cannot give a precise indication of the pH, only a range.

▶ Most indicators are polar solutions so cannot be used to identify the pH in non-polar solutions.

5 a In an acidic solution, the concentration of H_3O^+ increases. This increase will drive the reaction forward in order to remove excess H_3O^+ and, in turn, increase the concentration of bromothymol yellow, changing the colour of the indicator.

  b bromothymol blue $\rightleftharpoons$ bromothymol yellow $+ OH^-$

  c In an alkali solution, the concentration of OH^- increases. This increase will shift the reaction to the left in order to remove excess OH^- and in turn increase the concentration of bromothymol blue, changing the colour of the indicator.

WS 5.2 PAGE 55

1 a $SO_2(g) + CaO(s) \rightarrow CaSO_3(s)$

  b $n(SO_4) = \dfrac{140}{24.79} = 5.65 \ moles$

  Ratio = 1 : 1; therefore, $n(CaO) = 5.65 \ moles$

  $m(CaO) = n \times MM = 5.65 \times (40.08 + 16.00) = 317 \ g$

2 a $NaHCO_3(s) + HCl(aq) \rightarrow NaCl(aq) + H_2O(l) + CO_2(g)$

  b $n(HCl) = c \times V = 22.0 \times 10.0 = 220 \ moles$

  Ratio = 1 : 1; therefore, $n(NaHCO_3) = 220 \ moles$

  $m(NaHCO_3) = n \times MM = 220 \times (22.99 + 1.008 + 12.01 + (16.00 \times 3))$

  $= 18\,481.76 \ g = 18.5 \ kg$

3 a $NaOH(s) + HCl(aq) \rightarrow NaCl(aq) + H_2O(l)$

  $n(NaOH) = \dfrac{m}{MM}$

  $= \dfrac{200}{22.99 + 16.00 + 1.008}$

  $= 5.00 \ moles$

  $V(HCl) = \dfrac{5.00}{6} = 0.8 \ L$

  b The use of $6 \ mol \ L^{-1}$ HCl is a poor choice to neutralise this spill as it is a concentrated acid and poses safety concerns due to it being highly corrosive. The reaction between HCl and NaOH is exothermic, which could cause the heated water to spit, potentially causing burns.

4 a $KOH(aq) + HF(aq) \rightarrow KF(aq) + H_2O(l)$

  b $n(KOH) = c \times V = 1.2 \times 0.030 = 0.036 \ moles$

  c $q = mC\Delta T$

  $= 0.07 \times 4.18 \times 10^3 \times 5.7 = 1667.82 \ J$ or $1.7 \ kJ$ (2 sig fig)

  d Ratio $KOH : H_2O = 1 : 1$; therefore, $n(H_2O) = 0.036 \ moles$

  e $\Delta H_{neut} = \dfrac{q}{n(H_2O)} = \dfrac{1.67}{0.036} = 46.4 \ kJ \ mol^{-1}$

5 a $HCl(aq) + NaOH(aq) \rightarrow NaCl(aq) + H_2O(l)$

  b $T_{initial} = 21.0°C$

  $T_{final} = 21.7°C$ (highest temperature reached)

  c The highest temperature was 21.7°C, which occurred when 12 mL of HCl was added. Even though more HCl was added, the temperature in the calorimeter did not increase further as all the sodium hydroxide had been neutralised. No further bonds were being broken, therefore, no more energy was being released into the solution.

  d $q = mC\Delta T = 0.0168 \times 4.18 \times 10^3 \times 0.7 = 49.157 \ J$

  $n(NaOH) = c \times V = 0.25 \times 0.0048 = 0.0012 \ moles$

  $n(NaOH) = n(H_2O)$; therefore, $n(H_2O) = 0.0012 \ moles$.

  $\Delta H_{neut} = \dfrac{q}{n(H_2O)} = \dfrac{49.157}{0.0012} = -40\,964 \ J$ or $-41 \ kJ \ mol^{-1}$

  e ▶ Heat lost to surroundings

    ▶ Error when creating the concentration of solutions

    ▶ Parallax error when reading the thermometer

1 A   **2** B   **3** D   **4** D

5

	Arrhenius	**Brønsted–Lowry**
Acid	**A substance that ionises in solution to produce hydrogen ions**	**A proton donor**
Base	**A substance that in solution produces hydroxide ions**	**A proton acceptor**
Example chemical equation	$HCl(aq) \rightarrow H^+(aq) + Cl^-(aq)$	$HCl(aq) + H_2O(l) \rightarrow H_3O^+(aq) + Cl^-(aq)$

6 Lavosier proposed that all oxides dissolve in water to produce acidic solutions and all acids contain oxygen.

This definition is limited as it failed to explain the nature of some acids, such as hydrochloric acid, that do not contain oxygen. It also failed to explain oxides, such as magnesium oxides, that produce bases that dissolve in water.

Davy worked with other hydrohalic acids, such as HCl, and redefined the definition of acids as containing hydrogen, rather than oxygen. However, Davy's definition did not explain why compounds such as methane (CH_4) are not acidic, nor did it explain bases.

To overcome this, Arrhenius suggested that rather than hydrogen atoms, when dissolved in water all acids ionise to form hydrogen ions and that strong acids are completely ionised in water, whereas weak acids are only partially ionised. Expanding the definition to include bases, when dissolved in water, hydroxide ions are formed. All acid–base reactions involve hydrogen ions reacting with hydroxide ions to form water. Although this solved many of the previous problems with the definition, Arrhenius' definition restricted the definition to aqueous solutions. It failed to explain compounds such as ammonia that are basic, rather than acidic, and salts such as zinc chloride and sodium sulfide that are acidic and basic but do not contain hydrogen or hydroxide ions.

Brønsted and Lowry redefined acids as proton donors and bases as proton acceptors. Acids give up a proton and become a conjugate base, whereas bases accept the proton and become a conjugate acid. This definition is still limited as it does not explain the reactions of acidic oxides with bases or basic oxides with acids. It also cannot explain the acidic nature of chlorides that do not have a proton to donate.

~continue in right column ▲

7 The Brønsted–Lowry definition expands on the Arrhenius definition and does not require reactions to take place within a solvent – if a proton donor and proton acceptor are still present an acid–base reaction can occur. As most students learn about acids and bases in solutions, Arrhenius' definition allows for a simple understanding of how salt and water forms during neutralisation.

Chapter 6: Using Brønsted–Lowry theory

WS 6.1 PAGE 60

1

2 B. NH_4Cl is an acidic salt; H_2SO_4 is a strong acid; NaBr is neutral.

3 A. $1 \times 10^{-1.74} = 0.018 \text{ mol}^{-1}$

4 A. $Ba(OH)_2[OH^-] = 4.5 \times 10^{-3} \times 2 = 0.009$
$-\log[0.009] = 2.0$; therefore, pH = 14 – 2.0 = 12

5 D. As a log calculation, normal sig fig rules don't apply. The number of decimal places in the concentration becomes the sig figs of the pH.

6 D. $[H^+] = 10^{-4}$; therefore, $[OH^-] = \dfrac{10^{-14}}{10^{-4}}$ as $[H^+][OH^-] = 10^{-14} \text{ mol}^{-1}$

7

	$[H^+]$ (mol L^{-1})	pOH	$[OH^-]$ (mol L^{-1})
pH of orange juice = 3.1	$[H^+] = 10^{-3.1}$ $= 7.9 \times 10^{-4}$	14 – 3.1 = 10.9	$[OH^-] = 10^{-10.9}$ $= 1.3 \times 10^{-11}$
pH of kombucha = 2.6	$[H^+] = 10^{-2.6}$ $= 2.5 \times 10^{-3}$	14 – 2.6 = 11.4	$[OH^-] = 10^{-11.4}$ $= 4.0 \times 10^{-12}$
pH of milk = 6.8	$[H^+] = 10^{-6.8}$ $= 1.6 \times 10^{-7}$	14 – 6.8 = 7.2	$[OH^-] = 10^{-7.2}$ $= 6.3 \times 10^{-8}$
pH of human blood = 7.8	$[H^+] = 10^{-7.8}$ $= 1.6 \times 10^{-8}$	14 – 7.8 = 6.2	$[OH^-] = 10^{-6.2}$ $= 6.3 \times 10^{-7}$
pH of oven cleaner = 13.2	$[H^+] = 10^{-13.2}$ $= 6.3 \times 10^{-14}$	14 – 13.2 = 0.8	$[OH^-] = 10^{-0.8}$ $= 1.6 \times 10^{-1}$

8 a The calculation is correct as $-\log(1.2) = -0.079$. A negative pH is possible. Strong acids with a concentration greater than 1 will have a negative pH.

b The student could measure the pH of the solution by adding 50 mL of the solution to a beaker and recording the pH using a probe (an indicator would not be accurate to provide a negative numerical value).

9 a $CO_2(aq) + H_2O(l) \rightleftharpoons H_2CO_3(aq)$

b 2000: pH = 8.07
$[H^+] = 1 \times 10^{-8.07} = 8.51 \times 10^{-9}\,mol\,L^{-1}$
2020: pH = 8.03
$[H^+] = 1 \times 10^{-8.03} = 9.33 \times 10^{-9}\,mol\,L^{-1}$

c pH = $-\log(8.13 \times 10^{-9}) = 8.09$; therefore, sample was taken in 1985.

d $12\,\mu\,mol\,kg^{-1}$

e $0.0000125\,mol\,kg = 12.5\,\mu\,mol\,kg^{-1} = pH\,8.07$
$[H^+] = 1 \times 10^{-8.07} = 8.51 \times 10^{-9}\,mol\,L^{-1}$

WS 6.2 PAGE 63

1 a This model is not useful as it incorrectly defines the terms. On the concentrated side, the model shows the purple dots close together and on the dilute side the purple dots are spread out. This implies concentrated means the atoms are 'close together' and dilute means they are 'far apart'. Concentration is a measure of the number of molecules in solution, not their grouping. In the model, strong is modelled with seven purple dots and weak with three purple dots. Strong and weak does not mean the number of molecules but rather the measure of the ionisation of the molecules.

b

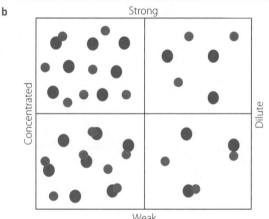

2 a
$NH_3 + H_2O \rightleftharpoons NH_4^+ + OH^-$

b
$HI + OH^- \rightleftharpoons I^- + H_2O$

c
$HNO_3 + HCO_3^- \rightleftharpoons H_2CO_3 + NO_3^-$

3 a

Acid/Base	Conjugate	Acid or base?
H_2O	OH^- H_3O^+	Base; acid
OH^-	H_2O	Acid
HCO_3^-	H_2CO_3 or CO_3^{2-}	Acid or base
H_3PO_4	$H_2PO_4^-$	Base

H_2S	HS^-	Base
HSO_4^-	H_2SO_4 or SO_4^{2-}	Acid or base
ClO_4^-	$HClO_4$	Acid
$HBrO_2$	BrO_2^-	Base

b H_2O, HCO_3^- and HSO_4^- can act as both acids and bases. When acting as an acid, H_2O, HCO_3^- and HSO_4^- can donate a proton to form the conjugate bases OH^-, CO_3^{2-} and SO_4^{2-}. Alternatively, they can accept protons, acting as a base, to become the conjugate acids H_3O^+, H_2CO_3 and H_2SO_4.

4 a $HNO_3(aq) + H_2O(l) \rightarrow H_3O^+(aq) + NO_3^-(aq)$
$HCOOH(aq) + H_2O(l) \rightleftharpoons H_3O^+(aq) + HCOO^-(aq)$

$C_3H_5O(COOH)_3(aq) + H_2O(l) \rightleftharpoons H_3O^+(aq) + (C_3H_5O(COOH)_2COO)^-(aq)$

$(C_3H_5O(COOH)_2COO)^-(aq) + H_2O(l) \rightleftharpoons H_3O^+(aq) + (C_3H_5O(COOH)(COO)_2)^{2-}(aq)$

$(C_3H_5O(COOH)(COO)_2)^{2-}(aq) + H_2O(l) \rightleftharpoons H_3O^+(aq) + (C_3H_5O(COO)_3)^{3-}(aq)$

b pH is the measure of hydronium ions in solution. The concentration of the acids needs to be the same to determine how the pH changes based on the strength of the acids, rather than its concentration.

c A pH probe provides an accurate numerical reading that allows you to distinguish between formic and citric acid. The use of an indicator would not provide a clear numerical value, but rather a range of possible pH values or a subjective judgement as to a pH value. As pH increases by a factor of 10, this would invalidate the measurements as you would not be able to compare the relationship between concentration and pH between monoprotic formic acid and triprotic citric acid.

d % nitric acid $= \dfrac{[H_3O^+]}{[HNO_3]} = \dfrac{[1 \times 10^1]}{[0.01]} \times 100 = 100\%$

% citric acid $= \dfrac{[H_3O^+]}{[C_3H_5O(COOH)_3]} = \dfrac{[1 \times 10^{-2.1}]}{[0.01]} \times 100 = 79\%$

% formic acid $= \dfrac{[H_3O^+]}{[HCOOH]} = \dfrac{[1 \times 10^{-2.3}]}{[0.01]} \times 100 = 50\%$

e Nitric acid is a monoprotic acid that ionises 100%; therefore, it is a strong acid.
Citric acid is a triprotic acid that ionises 79%, making it a weak acid.
Formic acid is a monoprotic acid with a 50% ionisation rate, making it the weakest acid tested.

f HNO_3 is a strong acid that will produce a weak conjugate base, NO_3^-. As this base is weak, the protons donated to water are bound too strongly for the weak base to recapture and so it remains ionised. On the other hand, a weak acid such as HCOOH will produce a strong conjugate base, $HCOO^-$, which can easily recapture the donated protons limiting the $[H^+]$ concentration.

WS 6.3 PAGE 66

1 a $[OH^-] = 0.01\,mol\,L^{-1}$ as KOH is 1:1

b $c_2 = \dfrac{c_1 V_1}{V_2} = \dfrac{0.01 \times 0.040}{0.1} = 0.004\,mol\,L^{-1}$

$[OH^-] = 0.004$; therefore, pOH $= -\log(0.004) = 2.4$
pH = 14 $-$ pOH = 14 $-$ 2.4 = 11.6

c Adding more water lowers the concentration of OH^- ions in the solution, therefore decreasing the pH closer to 7.

d A 100 factor dilution is the same as a decrease of 2 pH units; therefore, the new pH = 11.6 − 2 = 9.6, and therefore, pOH = 14 − 9.6 = 4.4 mol L^{-1}.

2 $[H^+] = 1 \times 10^{-2.1} = 0.0079\,\text{mol L}^{-1}$

$$V_2 = \frac{c_1 V_1}{c_2} = \frac{0.52 \times 0.02}{0.0079} = 1.3\,\text{L}$$

3 a pOH = 14 − pH = 14 − 12.6 = 1.4
$[OH^-] = 1 \times 10^{-1.4} = 0.0398\,\text{mol L}^{-1}$

b $HNO_3(aq) + NaOH(aq) \rightarrow NaNO_3(aq) + H_2O(l)$

c $[OH^-] = 0.0398\,\text{mol L}^{-1}$
Ratio $HNO_3 : NaOH = 1 : 1$; therefore, 0.0398 mol L^{-1} required.

$$V_2 = \frac{c_1 V_1}{c_2} = \frac{0.0398 \times 0.3}{0.4} = 0.03\,\text{L}$$

d The resulting solution will be acidic as only 30 mL was required to neutralise the NaOH, leaving 7 mL of HNO$_3$ in excess.

e 0.007 L of a 0.4 mol L^{-1} HNO$_3$ solution within 0.337 L

$$c_2 = \frac{c_1 V_1}{V_2} = \frac{0.4 \times 0.007}{0.337} = 0.0083\,\text{mol L}^{-1}$$

pH = −log(0.0083) = 2.08

4 $[H^+] = 1 \times 10^{-1.4} = 0.04\,\text{mol L}^{-1}$

Students may choose to use any starting and finishing volume. Sample answer provided.

$$V_2 = \frac{c_1 V_1}{c_2} = \frac{0.04 \times 0.05}{0.004} = 0.5\,\text{L}$$

Method:

1 Pipette 50 mL of perchloric acid into a 500 mL volumetric flask.

2 Fill the flask with distilled water to the 500 mL mark.

3 Stopper the flask.

4 Invert the flask to ensure even distribution.

5 a $2HCl(aq) + Ba(OH)_2(aq) \rightarrow BaCl_2(aq) + 2H_2O(l)$
$n(HCl) = c \times V = 0.015 \times 0.1 = 0.0015\,\text{moles}$
$n(Ba(OH)_2) = c \times V = 0.05 \times 0.1 = 0.005\,\text{moles}$

$\dfrac{n(HCl)}{2} = 0.00075$; therefore, HCl is limiting.

Remaining $n(Ba(OH)_2) = 0.005 − 0.00075 = 0.00425\,\text{moles}$ making the remaining solution basic.

$n(OH^+ \text{ ions in excess}) = 0.00425 \times 2 = 0.0085\,\text{moles}$

$$c = \frac{n}{V} = \frac{0.0085}{0.2} = 0.0425\,\text{mol L}^{-1}$$

pOH = −log(0.0425) = 1.4
pH = 14 − 1.4 = 12.6

b $c_2 = \dfrac{c_1 V_1}{V_2} = \dfrac{1 \times 10^{-12.6} \times 0.20}{0.350} = 1.44 \times 10^{-13}\,\text{mol L}^{-1}$

pH = −log(1.44 × 10^{-13}) = 12.8

6 $2NaOH(aq) + H_2SO_4(aq) \rightarrow Na_2SO_4(aq) + 2H_2O(l)$

$$n(NaOH) = \frac{m}{MM} = \frac{1.6}{22.99 + 16.00 + 1.008} = 0.040\,\text{moles}$$

$n(H_2SO_4) = c \times V = 0.28 \times 0.150 = 0.042\,\text{moles}$

Ratio $NaOH : H_2SO_4 = 2 : 1$; therefore, $\dfrac{0.04}{2} = 0.02\,\text{moles}$
= 0.02 moles of H$_2$SO$_4$ is required.

Excess $H_2SO_4 = 0.042 − 0.02 = 0.022\,\text{moles}$
As H$_2$SO$_4$ is diprotic, $n(H^+) = 0.022 \times 2 = 0.044\,\text{moles}$.

$$[H^+] = \frac{n}{V} = \frac{0.044}{0.050 + 0.150} = 0.22\,\text{mol L}^{-1}$$

pH = −log(0.22) = 0.66

---

7 a pH = −log(0.012) = 1.9. The solution will turn red.

b $c_2 = \dfrac{c_1 V_1}{V_2} = \dfrac{0.012 \times 0.008}{0.1} = 0.00096\,\text{mol L}^{-1}$

pH = −log(0.00096) = 3.0. The solution will turn violet.

c $HBr(aq) + KOH(aq) \rightarrow KBr(aq) + H_2O(l)$
$n = c \times V = 0.00096 \times 0.03 = 0.0000288\,\text{moles}$
Ratio HBr : KOH = 1 : 1; therefore, 0.0000288 moles is required.

$$V = \frac{n}{c} = \frac{0.0000288}{0.0040} = 7.2\,\text{mL}$$

WS 6.4 PAGE 69

1 a $K_a = \dfrac{[CH_3COO^-][H^+]}{[CH_3COOH]}$

b $1.8 \times 10^{-5} = \dfrac{[CH_3COO^-][H^+]}{[CH_3COOH]}$

Let x moles of CH$_3$COOH, which ionise.

$$1.8 \times 10^{-5} = \frac{[x][x]}{[CH_3COOH - x]} = \frac{[x]^2}{[0.30 - x]}$$

Assume x is much smaller than 0.30 mol L^{-1}
$x^2 = 1.8 \times 10^{-5} \times 0.30 = 5.4 \times 10^{-6}$
$[H^+] = x = \sqrt{5.4 \times 10^{-6}} = 0.00232379\,\text{mol L}^{-1}$
pH = −log(0.00232379) = 2.6338 = 2.6

2 a

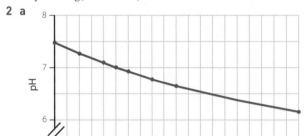

b 6.9

c As the temperature increases, the K_w increases. The dissociation of water molecules into ions is an endothermic process as energy must be absorbed to break the bonds.

3 a $HCOOH(aq) + H_2O(l) \rightleftharpoons HCOO^-(aq) + H_3O^+(aq)$

b $K_a = \dfrac{[HCOO^-][H_3O^+]}{[HCOOH]} = 1.77 \times 10^{-4}$

c $pK_a = -\log K_a = -\log(1.77 \times 10^{-4}) = 3.75$

d $[HCOOH] = 0.250\,\text{mol L}^{-1}$
Let x moles of HCOOH ionise.

$$pK_a = \frac{[HCOO^-][H_3O^+]}{[HCOOH]} = 1.76 \times 10^{-5}$$

$$1.76 \times 10^{-5} = \frac{[x][x]}{[0.250 - x]}$$

Assume x is small by comparison with 0.250 mol L^{-1}.
$x^2 = 1.76 \times 10^{-5} \times 0.250 = 4.4 \times 10^{-6}$
$x = 2.1 \times 10^{-3}\,\text{mol L}^{-1} = [H_3O^+]$
pH = −log(2.1 × 10^{-3}) = 2.7

e Acetic acid has a smaller K_a, so it will ionise to a lower extent than formic acid, producing a lower concentration of H$_3$O$^+$, and therefore will have a higher pH.

4 a Acid Y is stronger as it has a lower pK_a value than HX. pK_a is the negative log of the acid dissociation constant or K_a value; therefore, a low pK_a value describes an acid that is more fully dissociated in the water.

b $K_a(HX) = 10^{-3.86} = 1.38 \times 10^{-4}$

$$K_b = \frac{K_w}{K_a} = \frac{1 \times 10^{-14}}{1.38 \times 10^{-4}} = 7.24 \times 10^{-11}$$

Therefore, the weaker the acid, the stronger the conjugate base.

c $K_a(HY) = 10^{-1.92} = 1.2 \times 10^{-2}$

$$K_a = \frac{[H^+][Y^-]}{[HY]}$$

Assuming x is small, therefore:

$$1.2 \times 10^{-2} = \frac{x^2}{0.230}$$

$$x^2 = 1.2 \times 10^{-2} \times 0.230 = 0.002765$$

$$x = \sqrt{0.002765}$$

Therefore, $[H^+] = 0.0526 \, mol \, L^{-1}$

$pH = -\log(0.0526) = 1.28$

Chapter 7: Quantitative analysis

WS 7.1 PAGE 72

1 a Did not ensure Na_2CO_3 was dry before weighing; used tap water, not distilled water; filled volumetric flask before adding Na_2CO_3; the bottom, rather than the top, of the meniscus should be on the mark; the flask should be inverted several times rather than shaken.

b Dry Na_2CO_3 in the oven until constant mass is reached. Accurately weigh required amount using an electronic balance, then dissolve it in distilled water. Transfer it to a volumetric flask using a funnel, rinsing the reaction vessel and funnel with distilled water. Top up the flask to the mark with distilled water so that the bottom of the meniscus is on the mark. Stopper the flask and invert several times to mix.

c ▸ If the Na_2CO_3 was not dry, then the calculated number of moles would be higher than the actual number of moles so the calculated concentration would be higher than the actual concentration.

▸ Filling the volumetric flask before adding the sodium carbonate means there is a greater volume of water than indicated by the volumetric flask volume so the calculated concentration would be higher than the actual concentration.

▸ There may be sodium and/or carbonate ions in the tap water, which would increase the actual concentration of these ions compared to the calculated concentration.

2 a Large molecular mass is important to ensure accurate mass is weighed.

High purity and unreactive to air or moisture is important as the concentration must be accurate and not contaminated.

b $n(KH(C_8H_4O_4)) = c \times V = 0.020 \times 0.150 = 0.0030$ moles
$m(KH(C_8H_4O_4)) = n \times MM = 0.0030 \times 204.2 = 0.61226$
$\qquad = 0.61 \, g$

3 a The pipette should be rinsed with distilled water and then rinsed with the sodium carbonate solution prior to being filled for use with the sodium carbonate solution. This is done as any remaining water in the pipette would dilute the sodium carbonate solution, so the actual concentration would be less than the calculated concentration, compromising the concentration of the sodium carbonate.

b The burette should be rinsed with distilled water and then rinsed with the hydrochloric acid solution prior to being filled for use with the hydrochloric acid solution. This is done as any remaining water in the burette would dilute the acid solution, so its concentration would be less than the calculated concentration.

4 a Sulfuric acid can cause severe skin burns, can irritate the nose and throat and cause difficulties breathing if inhaled, can burn the eyes, possibly causing blindness. The student should wear safety glasses and perform this experiment in a well-ventilated room. If contact with skin occurs, rinse immediately with water and seek medical attention.

b $H_2SO_4 + 2NaOH \rightarrow Na_2SO_4 + 2H_2O$
$n(NaOH) = c \times V = 1.16 \times 0.0171 = 0.01984$ moles
Ratio $NaOH : H_2SO_4 = 2 : 1$; therefore, $n(H_2SO_4) = \frac{0.01984}{2} = 0.00992$ moles.

$$c = \frac{n}{V} = \frac{0.00992}{0.002} = 4.96 \, mol \, L^{-1}$$

c The calculated concentration is slightly higher than the manufacturer's claims. To improve accuracy, the student should have standardised the NaOH prior to the titration. As the concentration may be lower than stated by the manufacturer, a higher volume of NaOH is required to neutralise the acid. To improve reliability, the titration should have been performed several times until the volume of the sodium hydroxide was measured consistently to within 0.05 mL, with the average volume used in the calculation.

5 a Phenolphthalein should be used as it will go from clear to a faint pink when the end point is reached. A weak acid and strong base will have an end point around pH 9.

b

Attempt	mL NaOH added
1	25.70
2	23.95
3	24.15
4	24.10
Average	24.07

c The first titration is often called a rough titration and is used to roughly determine the end point before consecutive, more accurate and careful titrations are performed. It is not to be included in the calculation of the average.

d i $C_6H_8O_7(aq) + 3NaOH(aq) \rightarrow Na_3C_6H_5O_7(aq) + 3H_2O(l)$

ii $n(NaOH) = c \times V = 0.1045 \times 0.02407 = 0.002515$ moles
$NaOH : C_6H_8O_7 = 3 : 1$; therefore, $n(C_6H_8O_7) = \frac{n(NaOH)}{3} = \frac{0.002515}{3} = 0.0008384$ moles

$$\text{diluted}[C_6H_8O_7] = \frac{n}{V} = \frac{0.0008384}{0.025} = 0.03354 \, mol \, L^{-1}$$

$$\text{undiluted}[C_6H_8O_7] = 0.03354 \times \frac{200}{25} = 0.2683 = 0.268 \, mol \, L^{-1}$$

e Lemon juice contains other acids, such as malic acid. In the calculation, we are assuming that the lemon juice contained only citric acid. This assumption means the calculation is often higher than the actual concentration of citric acid present.

6 Moles of sodium carbonate used $= \frac{12.5}{1000} \times 0.0250 = 0.0003125$ moles

$Na_2CO_3(aq) + 2HCl(aq) \rightarrow 2NaCl(aq) + CO_2(g) + H_2O(l)$
Ratio $Na_2CO_3 : HCl = 1 : 2$; therefore, $n(HCl)$ in excess $= 2 \times 0.0003125 = 0.000625$ moles
Moles HCl added to flask to react with $NH_3 = \frac{70.00}{1000} \times 0.100 = 0.0070$ mol

Therefore, $n(\text{HCl})$ reacted with $NH_3 = 0.0070 - 0.000\,625 =$ 0.006375 moles

$NH_3(aq) + HCl(aq) \rightarrow NH_4Cl(aq)$

Therefore, $n(NH_3)$ in 20.0 mL samples = 0.006 375 moles

$[NH_3]$ in original $= \dfrac{1000}{20} \times 0.006\,375 = 0.31875 = 0.319\,\text{mol}\,L^{-1}$

7 a Standardisation of NaOH

Preparation of standard solution:

1 Solid oxalic acid dihydrate was carefully weighed in a beaker.

2 A small volume of distilled water was added to dissolve the solid.

3 The funnel and beaker were rinsed with distilled water, which was added to the flask. More distilled water was

~continue in right column ▲

then added to the flask until the bottom of the meniscus hit the mark

Titration:

1 Clean and rinse the burette with sodium hydroxide solution. Using a funnel, fill the burette with sodium hydroxide solution and record the initial volume.

2 Rinse the pipette with oxalic acid standard solution then pipette 25 mL aliquots of oxalic acid solution into four clean 100 mL conical flasks. Add three drops of phenolphthalein to each flask.

3 Titrate the sodium hydroxide solution against the oxalic acid solution until a faint pink colour is seen. Record the volume of sodium hydroxide used.

b

Step	Justification
Refill the burette with sodium hydroxide solution. Record the initial volume.	Sodium hydroxide was put in the burette in both titrations to minimise the need for cleaning and rinsing of equipment. It ensures the concentration of sodium hydroxide is maintained and not diluted by oxalic acid residue.
Pipette 25 mL aliquots of the diluted kiwi fruit juice into four clean 100 mL conical flasks. Add three drops of phenolphthalein to each flask.	Conical flasks must be clean and free of residue to ensure the colour change observed is due to the neutralisation of the kiwi fruit juice and not other contaminants. Phenolphthalein is a suitable indicator for this weak acid–strong base titration. Four conical flasks are used in order to perform four titrations to obtain an average for reliability.
Titrate the sodium hydroxide against the diluted kiwi fruit juice until a faint pink colour is seen. Record the volume of sodium hydroxide used.	The titration must stop when a faint pink colour remains after agitation of the conical flask. A deeper pink colour means the student has overshot the end point.

c Answers may include:

▶ The electronic balance used to weigh the oxalic acid dihydrate for the primary standard may not have been calibrated properly.

▶ The top of the meniscus was always on top of the calibration mark, rather than the bottom.

▶ Students chose an endmark colour that was too dark.

d If the mass of oxalic acid dihydrate is incorrect, all concentration calculations performed in the investigation would also be incorrect, and the final calculation of the concentration of ascorbic acid within the kiwi fruit would be wrong. If the top of the meniscus was read rather than the bottom, the titration value would be less so the calculated concentration would be lower than the actual concentration. If the endmark colour was darker, then too much sodium hydroxide would have been added so the calculated concentration would be higher than the actual concentration.

e The electronic balance should be calibrated prior to use, using a set of scale calibration weights, and all solid must be transferred to the volumetric flask. The bottom of the meniscus should be read. The endmark colour should be a light pink and the colour of all trials should be compared for consistency.

f Average volume of NaOH = 11.66 mL

Attempts 1 and 3 were excluded as they exceeded a 0.1 mL difference from the other attempts.

g $n(\text{NaOH}) = c \times V = 0.100 \times 0.011\,66 = 0.001\,166\,\text{moles}$

$H_2C_6H_6O_6(aq) + 2NaOH(aq) \rightarrow 2H_2O(l) + Na_2C_6H_6O_6(aq)$

Ascorbic acid is diprotic, therefore, ratio =

$HC_6H_6O_6 : NaOH = 1 : 2$

$n(H_2C_6H_6O_6) = \dfrac{0.001166}{2} = 0.000\,583\,\text{moles}$

$c(H_2C_6H_6O_6)$ in diluted sample $= \dfrac{n}{V} = \dfrac{0.000\,583}{0.025}$
$= 0.023\,32\,\text{mol}\,L^{-1}$

$c(H_2C_6H_6O_6)$ in original sample $= 0.023\,32 \times \dfrac{100}{20}$
$= 0.1166 = 0.117\,\text{mol}\,L^{-1}$

WS 7.2 PAGE 79

1 B. The starting pH of Acid 1 is higher than Acid 2, whereas their equivalence points are the same.

2 D. Methyl orange has a pH range of 3.4–4.4, which matches the equivalence point.

3 D. The end point is reached at a pH of 9.

4 B. Conductivity starts high due to the presence of highly mobile hydrogen ions and decreases as they react with OH^- to form water. Unlike strong acid–base titrations, the continued addition of weak base does not lead to an increase in conductivity.

5 C. pH of acid is above 5, pH of base starts at 14. This indicates a weak acid with a strong base.

6 a, b

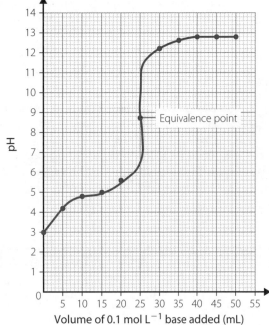

Volume of 0.1 mol L^{-1} base added (mL)

c The pH begins at 3, therefore, the acid is weak. Suitable weak acids could be acetic acid and formic acid.

The pH curve ends at 12.6, therefore, it is a strong base. Suitable strong bases could be sodium hydroxide or potassium hydroxide.

d 25 mL

e $H^+(aq) + OH^-(aq) \rightarrow H_2O(l)$

$n(OH^-) = c \times V = 0.1 \times 0.025 = 0.0025$ moles

Ratio H+ : OH$^-$ = 1 : 1; therefore, $n(H^+) = 0.0025$ moles

$c(H^+) = \dfrac{n}{V} = \dfrac{0.0025}{0.05} = 0.05 \, \text{mol L}^{-1}$

7 Carbonic acid is a diprotic acid.

$H_2CO_3(aq) + OH^-(aq) \rightleftharpoons H_2O(l) + HCO_3^-(aq)$

$HCO_3^-(aq) + OH^-(aq) \rightleftharpoons H_2O(l) + CO_3^{2-}(aq)$

The hydronium ions do not dissociate at the same time; therefore, the curve will produce an equivalence point for each ionisation.

8 a, b

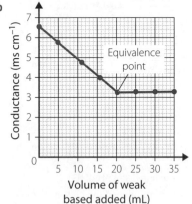

Volume of weak
based added (mL)

c Strong acids fully ionise in water, therefore the conductivity of the solution begins high. When the weak base is added, the hydrogen ion concentration decreases as the acid is neutralised. The decrease in conductivity continues until the acid is completely neutralised at the equivalence point. As more base is added, the conductivity remains unchanged as weak bases do not completely ionise in solution so the hydronium concentration remains low.

- -

WS 7.3 PAGE 84

1 a Calcium carbonate is insoluble in water and cannot be directly titrated. Therefore, it is dissolved first in excess hydrochloric acid. The solution containing excess HCl will be titrated

against sodium hydroxide to determine the amount of HCl that reacted with the calcium carbonate, and therefore the concentration of calcium carbonate in the original sample.

b i They needed to ensure the HCl was in excess so that all the calcium carbonate had reacted. They also needed to make sure there was excess unreacted HCl that could be titrated.

ii They could use an indicator or pH meter to check that the solution was acidic after the reaction was complete.

iii Carbon dioxide gas is a product of the reaction between calcium carbonate and hydrochloric acid. If it remains dissolved in the solution over time, the carbon dioxide will react with the water to produce carbonic acid. This will increase the amount of sodium hydroxide required to reach the end point of the titration and lead to a much larger calcium carbonate percentage being calculated. $CO_2(g)$ is in equilibrium with $CO_2(aq)$. By heating the solution, it is favouring the formation of $CO_2(g)$.

c i Sodium hydroxide will react with air and absorb moisture from the environment, so although the original concentration may be mathematically accurate, the concentration of the solution used may be less. Sodium hydroxide also has a small molecular mass which may introduce errors when weighing.

ii Bromothymol blue is the best indicator to use in this titration as this titration is a strong acid–strong base titration that will have an equivalence point around pH 7. Bromothymol blue will turn green at the end point that corresponds to the equivalence point.

d $n(\text{NaOH})$ used $= \dfrac{26.65}{1000} \times 0.15 = 0.003\,9975$ moles

$\text{NaOH}(aq) + \text{HCl}(aq) \rightarrow \text{NaCl}(aq) + H_2O(l)$; therefore, NaOH : HCl = 1 : 1

$n(\text{HCl})$ in 25 mL = 0.003 9975 moles

$n(\text{HCl})$ in 250 mL = $0.003\,9975 \times \dfrac{250}{25} = 0.039975$ moles

$n(\text{HCl})$ added to flask to react with limestone $= \dfrac{100.0}{1000} \times 0.85 = 0.085$ moles

Therefore, $n(\text{HCl})$ reacted with limestone = 0.085 – 0.039 975 = 0.045025 moles

$CaCO_3(s) + 2HCl(aq) \rightarrow CaCl_2(aq) + CO_2(g) + H_2O(l)$

Ratio $CaCO_3$: HCl = 1 : 2; therefore, $n(CaCO_3) = \dfrac{0.045\,025}{2} = 0.022\,51...$ moles

$m(CaCO_3) = n \times MM = 0.0225 \times 100.1 = 2.2532 \, \text{g}$

Therefore, %$CaCO_3$ in limestone $= \dfrac{2.25}{5.0} \times 100 = 45.06\% = 45\%$

2 a

Chemical process	Procedure step
The protein found in cereal can be oxidised to form the ammonium ion.	Steps 2–3
The ammonium ion can be converted into ammonia.	Step 5
The amount of unreacted acid is determined by back titration.	Step 7

b The ammonium ion is produced by using the strong acid H_2SO_4. As the ammonium ion is a weak acid, titrating this solution would also include the concentration of sulfuric acid within the solution and the amount of ammonium ion would not be able to be determined.

c If any of the ammonia evaporates before it can be added to the HCl solution, the final calculation of protein within the cereal will be incorrect as loss of ammonia will result in a lower calculation of protein present.

WS 7.4 PAGE 87

1 a A buffer is a solution that resists change in pH because it contains comparable amounts of a weak acid and its **conjugate base**.

b The natural buffering that occurs in some lakes is a result of the dissolution of **carbon dioxide** from the air.

c The other half of a buffer system containing sodium acetate would be **acetic acid**.

d The CO_2 dissolves in blood (water) to produce H_2CO_3.

e The other part of the buffer system is the HCO_3^- ion.

f If the concentration of OH^- ions increase in the blood, H_2CO_3 reacts to **reduce** the OH^- concentration.

2 Buffering capacity is the amount of acid or base that can be neutralised by a buffer system. It is determined by the concentrations of the acid and conjugate base. If a volume of acid or base is added that exceeds the buffer's capacity, the pH will quickly change.

3 Buffers with equal concentrations of acid and conjugate base have the equal capacity to neutralise added acid or base. If the concentration of one was lower, it would lower the buffering capacity.

4 $H_2PO_4^- + H_2O(l) \rightarrow HPO_4^{2-}(aq) + H_3O^+(aq)$

5 When an acid is added to the system, the extra H^+ ions are consumed by the HPO_2^{2-}: $HPO_2^{2-} + H_3O \rightleftharpoons H_2PO_4^- + H_2O$; therefore, forming a weaker acid and minimising change to the pH. When a base is added to the system, the extra OH^- reacts with the H_3O^+, hence decreasing the $[H_3O^+]$ and favouring the reverse reaction hence decreasing the $[H_2PO_4^-]$ and minimising the change in $[H_3O^+]$ and pH.

6 $H_3O(aq) + F^-(aq) \rightleftharpoons HF(aq) + H_2O(l)$
The added OH^- ions will react with the H_3O^+ ions, reducing their concentration in the equilibrium mixture. This will force the reaction to the left to increase the $[H_3O^+]$, thus minimising the change in pH.

7 a $H_2PO_4^-$ is the acid, HPO_4^{2-} is the conjugate base.

b $K_{a2} = 6.2 \times 10^{-8}$ The second ionisation of phosphoric acid is $H_2PO_4^-$ to HPO_4^{2-}, which represents the conjugate acid–base pair in the reaction.

c $pK_a = -\log(6.2 \times 10^{-8}) = 7.2$
As concentrations of acid and base are equal, $pH = pK_a = 7.2$.

MODULE SIX: CHECKING UNDERSTANDING PAGE 89

| **1** A | **2** B | **3** D | **4** C | **5** C |
| **6** A | **7** B | **8** A | **9** C | **10** A |

11 a The Brønsted–Lowry theory overcame the limitations of Davy and Arrhenius to define acids as proton donors and bases as proton acceptors. This explained why some compounds that contained hydrogen atoms do not act as acids, such as ammonia.

b

Compound	Arrhenius	Brønsted–Lowry
Ethane (C_2H_6)	Ethane is not an acid as it does not dissociate to form H^+ in water.	Ethane is not an acid as it does not donate protons.
Ethanoic acid (CH_3COOH)	Ethanoic acid contains three hydrogen atoms but only one dissociates to form H^+ in water.	CH_3COOH can donate a proton to water to form the conjugate base CH_3COO^-.

Compound	Arrhenius	Brønsted–Lowry
Ammonia (NH_3)	Ammonia is not an acid as it does not dissociate to form H^+ in water.	Although more commonly a base, NH_3 can donate a proton to form NH_2^-.

12 a $n(H_2SO_4) = c \times V = 0.124 \times 0.050 = 0.0062$ mol
$n(LiOH) = c \times V = 0.200 \times 0.060 = 0.012$ mol
$H_2SO_4(aq) + 2LiOH(aq) \rightleftharpoons Li_2SO_4(aq) + 2H_2O(l)$
Ratio $H_2SO_4 : LiOH = 1 : 2$; therefore, LiOH is limiting and
$n(H_2SO_4) = \dfrac{0.012}{2} = 0.006$ moles.
As all the LiOH is neutralised, there remains $n(H_2SO_4) = 0.0062 - 0.006 = 0.0002$ moles in solution.
The resultant solution would be acidic.

b $c(H_2SO_4) = \dfrac{c}{V} = \dfrac{0.002}{0.110} = 0.00182\,\text{mol L}^{-1}$
Ratio $(H^+) : (SO_4^{2-}) = 2 : 1$; therefore, $c(H^+) = 2 \times 0.00182 = 0.00364$ moles
$pH = -\log[H] = -\log(0.00364) = 2.44$

c 1 Place the cabbage leaves in a large pot of boiling water.
2 Boil the cabbage leaves for 30 minutes and allow the liquid to cool.
3 Pour the liquid through a sieve, discarding the solid pieces of cabbage.
The cabbage will turn from purple to red in the presence of a strong acid.

d To determine pH, a correctly calibrated pH probe should be used rather than a natural indicator or other colourimetric indicator.

13 a $CH_3COOH(aq) + NaOH(aq) \rightarrow CH_3COONa(aq) + H_2O(l)$
$n(CH_3COOH) = c \times V = 0.365 \times 0.020 = 0.0073$ moles
Ratio $CH_3COOH : NaOH = 1 : 1$; therefore, $n(NaOH) = 0.00730$ moles
$c(NaOH) = \dfrac{n}{V} = \dfrac{0.00730}{0.0125} = 0.584\,\text{mol L}^{-1}$

b Analyte = sodium hydroxide; titrant = acetic acid

c Any four of the following responses:

Step	Justification
The burette was rinsed with sodium hydroxide	To ensure the concentration of the sodium hydroxide was unaltered
The pipette was rinsed with acetic acid	To ensure the concentration of the acetic acid was unaltered
The conical flask was rinsed with distilled water	To ensure that the flask was clean and that no excess sodium hydroxide or acetic acid was present
Phenolphthalein indicator was used	To match the end point with the equivalence point for the titration
Titration was repeated at least three times	To calculate an average and increase the reliability of results
White paper was placed under the conical flask	To easily determine the colour change

14 a

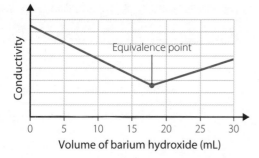

Conductivity vs Volume of barium hydroxide (mL), with "Equivalence point" marked around 18 mL.

b $V(KOH)$ at equivalence $= 18\,mL$

$n(H_2SO_4) = c \times V = 2.215 \times 10^{-4} \times 0.200 = 4.43 \times 10^{-5}$ moles

$H_2SO_4(aq) + 2KOH(aq) \rightarrow 2H_2O(l) + K_2SO_4(aq)$

Therefore, $n(KOH) = n(H_2SO_4) \times 2 = 8.86 \times 10^{-5}$ moles

$c(KOH) = \dfrac{n}{V} = \dfrac{8.86 \times 10^{-5}}{0.018} = 4.92 \times 10^{-3}\,mol\,L^{-1}$

c At the equivalence point, conductivity is lowest because the solution contains only potassium and sulfate ions. The conductivity then rises as K^+ and OH^- ions are added in excess. Although still highly mobile, OH^- ions are not as conductive as H^+; therefore, the gradient of the slope does not increase as steeply.

15 a $HCOOH + H_2O(l) \rightleftharpoons HCOO^-(aq) + H_3O^+(aq)$

In accordance with Le Chatelier's principle, if OH^- ions are added to the solution, the equilibrium will shift to the right to make up for the loss of hydrogen ions in the reaction with the base. If H^+ ions are added, the reaction will shift to the left, as H^+ ions react with $HCOO^-$ ions. In either case, the colour of the bromothymol blue will not change, as the pH would not significantly change with such small volumes of the acid and base being added.

b $K_a = \dfrac{[HCOO^-][H^+]}{[HCOOH]}$

c $1.77 \times 10^{-4} = \dfrac{[HCOO^-][H^+]}{[HCOOH]}$

Let x = number of moles of HCOOH that ionise.

$1.77 \times 10^{-4} = \dfrac{[x][x]}{[HCOOH - x]} = \dfrac{[x]^2}{[0.30 - x]}$

Assume x is much smaller than $0.30\,mol\,L^{-1}$

$x^2 = 1.77 \times 10^{-4} \times 0.30 = 5.31 \times 10^{-5}$

$[H^+] = x = \sqrt{5.31 \times 10^{-5}} = 0.007\,286\,97\,mol\,L^-$

$pH = -\log(0.007\,286\,97) = 2.1$

MODULE SEVEN: ORGANIC CHEMISTRY

REVIEWING PRIOR KNOWLEDGE PAGE 94

1 **a** False. Valency is the combining power of an element.
 b False. Carbon has a valency of 4.
 c True
 d False. Oxygen has a valency of 2.
 e False. Nitrogen has a valency of 3.
 f True
 g False. Combustion reactions are exothermic.
 h False. Hydrocarbons contain carbon and hydrogen only.
 i False. The formula for glucose is $C_6H_{12}O_6$.
 j True
 k True
 l False. Crude oil is a mixture.
 m True
 n True
 o True

2 **a** Methane

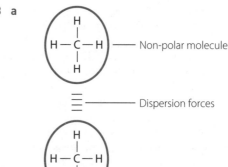

 b Ethanol

 c Acetic/ethanoic acid

3 **a** Non-polar molecule — Dispersion forces

 b Hydrogen bonds — Non-polar section — Dispersion forces — Non-polar section

4 **a** i $C_8H_{18}(l) + 12.5O_2(g) \rightarrow 8CO_2(g) + 9H_2O(g)$
  ii $C_2H_5OH(l) + 3O_2(g) \rightarrow 2CO_2(g) + 3H_2O(g)$
 b Octane is more likely to undergo incomplete combustion because it requires 12.5 moles of oxygen per mole of octane, which is more than four times the amount for ethanol, which requires 3 moles of oxygen per mole ethanol.
 c i Yellow
  ii There was incomplete combustion of methane, which produced the yellow flame and the soot, i.e. carbon that deposited on the bottom of the beaker. The equation is
  $2CH_4(g) + 2.5O_2(g) \rightarrow C(s) + CO(g) + 4H_2O(g)$

5 **a** $Y(s) \rightarrow Y^{2+}(aq) + 2e^-$
 b $Cr_2O_7^{2-}(aq) + 14H^+(aq) + 6e^- \rightarrow 2Cr^{3+}(aq) + 7H_2O(g)$
 c $3Y(s) + Cr_2O_7^{2-}(aq) + 14H^+(aq) \rightarrow 3Y^{2+}(aq) +$
  $2Cr^{3+}(aq) + 7H_2O(g)$
 d Oxidant: dichromate ions
  Reductant: Y solid

6 $1\,367\,000 = \dfrac{155 \times 4.18 \times 35}{n(\text{ethanol})}$

9780170449656

$$n(\text{ethanol}) = \frac{155 \times 4.18 \times 35}{1\,367\,000} = 0.016\,58\ldots$$

$MM(\text{ethanol, } C_2H_5OH) = (2 \times 12.01) + (6 \times 1.008) + 16.00 = 46.068\,\text{g mol}^{-1}$

$m(\text{ethanol}) = n \times MM = 0.01658\ldots \times 46.068\,\text{g} = 0.764\,\text{g (to 3 sig fig)}$

Chapter 8: Nomenclature

WS 8.1 PAGE 97

1 Hydrocarbons contain only the elements **carbon** and **hydrogen**. When compounds form from these elements with straight or branched chains, they are referred to as **aliphatic** hydrocarbons. These compounds can be saturated, i.e. they have only **single** bonds, or can be unsaturated, which means they have **double** or **triple** bonds. When hydrocarbons form ring structures, they are referred to as **alicyclic** hydrocarbons. The simplest member of this series is **cyclopropane**, which has a molecular formula of C_3H_6. These compounds can be saturated or **unsaturated**. An example of such an **alicyclic** compound that is used in school laboratories is **cyclohexene**, C_6H_{10}. Benzene, C_6H_6, belongs to a group called **aromatic** hydrocarbons.

~continue in right column ▲

2

Condensed formula	Alkane/alkene/alkyne	Saturated/unsaturated
CH_3CHCH_2	Alkene	Unsaturated
$CHCCH_2CH_3$	Alkyne	Unsaturated
CH_3CH_3	Alkane	Saturated
$CH_2CHCH_2CH_3$	Alkene	Unsaturated
$CH_3CH_2CH_3$	Alkane	Saturated
CH_3CCCH_3	Alkyne	Unsaturated

3 a Butane
  b Pentane
  c 3-Ethyl-4-methylhexane
  d 2-Bromo-4-iodopentane
  e But-2-ene
  f 3,4,6-Trimethyloctane
  g 3,4-Dimethylpent-2-ene
  h 1,1,1-Trichloro-2,2,2-trifluoroethane

4

	IUPAC name	Structural formula
a	2,2-Dimethylpropane	
b	Methylpropene	
c	3-Ethyl-3-methylheptane	
d	2,3-Dichloropent-2-ene	
e	Tetrachloromethane	
f	3-Fluoro-1,1,2-tribromopent-1-ene	

	IUPAC name	Structural formula
g	1,3,3-Tribromo-4-chlorobut-1-yne	$Br-C{\equiv}C-\overset{\displaystyle Br}{\underset{\displaystyle Br}{C}}-\overset{\displaystyle Cl}{\underset{\displaystyle H}{C}}-H$
h	3,4,4,5-Tetramethylheptane	

WS 8.2 **PAGE 100**

1	Structural formula	Condensed formula	IUPAC name
a		CH_3CH_2OH	Ethanol
b		$(CH_3)_3COH$	2-Methylpropan-2-ol
c		CH_3CH_2CHO	Propanal
d		$(CH_3CH_2)_2CO$	Pentan-3-one
e		HCOOH	Methanoic acid
f		$CH_3CH_2NH_2$	Ethanamine
g		$CH_3CH_2CONH_2$	Propanamide

2　**a**　2-Chloroprop-1-ene

　　b　4-Chloro-3-hydroxyheptanoic acid

　　c　2,2-Dimethylpent-4-enal

　　d　2-Aminobutanoic acid

　　e　1,1-Dichloro-5-hydroxy-pentan-3-one

　　f　4-Amino-2-bromo-3-methylpentanoic acid

　　g　3-Hydroxybutanal

　　h　2,2,3,3-Tetrafluoropropanoic acid

WS 8.3 **PAGE 102**

1　Structural isomers of C_5H_{12}

a	H H H H H ⎮ ⎮ ⎮ ⎮ ⎮ H—C—C—C—C—C—H ⎮ ⎮ ⎮ ⎮ ⎮ H H H H H	Pentane
b	H H H H ⎮ ⎮ ⎮ ⎮ H—C—C—C—C—H ⎮ ⎮ ⎮ ⎮ H ⎮ H H H—C—H ⎮ H	2-Methylbutane
c	H ⎮ H-C-H H ⎮ H ⎮ ⎮ ⎮ H—C—C—C—H ⎮ ⎮ ⎮ H ⎮ H H-C-H ⎮ H	2,2-Dimethylpropane

2　Possible isomers of C_6H_{12}

H_3C＼　　　／CH_3
　　　C＝C
H_3C／　　　＼CH_3
2,3-Dimethylbut-2-ene

H H H H H H
⎮ ⎮ ⎮ ⎮ ⎮ ⎮
H—C—C—C—C—C＝C
⎮ ⎮ ⎮ ⎮ ⎮ ⎮
H H H H H
Hex-1-ene

H H H H H H
⎮ ⎮ ⎮ ⎮ ⎮ ⎮
H—C—C—C—C＝C—C—H
⎮ ⎮ ⎮ ⎮ ⎮
H H H H H
Hex-2-ene

H H H H H H
⎮ ⎮ ⎮ ⎮ ⎮ ⎮
H—C—C—C＝C—C—C—H
⎮ ⎮ ⎮ ⎮
H H H H
Hex-3-ene

Cyclohexane

Can also include:

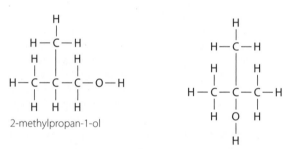

cis-Hex-2-ene　　　　trans-Hex-2-ene

cis-Hex-3-ene　　　　trans-Hex-3-ene

Other options include:　　3-methylpent-2-ene

2-methylpent-1-ene　　　4-methylpent-2-ene

3-methylpent-1-ene　　　2,3-dimethylbut-1-ene

4-methylpent-1-ene　　　3,3-dimethylbut-1-ene

2-methylpent-2-ene

3　**a**　Isomers with molecular formula $C_4H_{10}O$

H H H H
⎮ ⎮ ⎮ ⎮
H—C—C—C—C—O—H
⎮ ⎮ ⎮ ⎮
H H H H
butan-1-ol

H H H H
⎮ ⎮ ⎮ ⎮
H—C—C—C—C—H
⎮ ⎮ ⎮ ⎮
H H O H
⎮
H
butan-2-ol

H
⎮
H—C—H
H ⎮ H
⎮ ⎮ ⎮
H—C—C—C—O—H
⎮ ⎮ ⎮
H H H
2-methylpropan-1-ol

H
⎮
H—C—H
H ⎮ H
⎮ ⎮ ⎮
H—C—C—C—H
⎮ ⎮ ⎮
H O H
⎮
H
2-methylpropan-2-ol

　　b　**i**　Butan-1-ol and butan-2-ol; 2-methylpropan-1-ol and
　　　　　　2-methylpropan-2-ol

　　　　ii　Butan-1-ol and butan-2-ol; 2-methylpropan-1-ol and
　　　　　　2-methylpropan-2-ol

4　Functional group isomers with molecular formula C_3H_6O

H H
⎮ ⎮
H—C—C—C＝O
⎮ ⎮ ＼
H H H
Propanal

H H H
⎮ ⎮ ⎮
H—C—C—C—H
⎮ ⎮⎮ ⎮
H O H
Propanone

5　Chain/position/functional group

　　a　Functional group

　　b　Chain

　　c　Position

　　d　Functional group

　　e　Chain

　　f　Functional group

　　g　Position

6　Aldehydes, carboxylic acids and amides

WS 9.1 PAGE 104

1 a

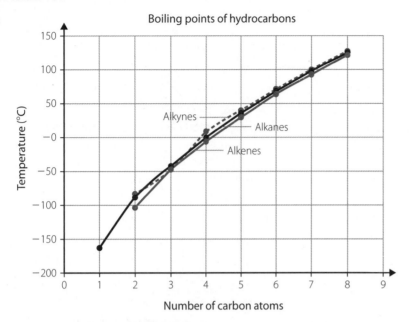

Boiling points of hydrocarbons

b All of these homologous series are non-polar hydrocarbons; hence, each have dispersion forces between the molecules of the various compounds. The boiling point increases as the number of carbon atoms increases within each homologous series. This is due to the increasing strength of dispersion forces between molecules. Straight chain alkynes and alkanes have higher boiling points than straight chain alkenes because of the stronger dispersion forces between their molecules due to their shape. This means that the molecules can align more closely than the alkenes, which have atoms fixed in a plane due to the fixed double bond. More energy is therefore required to separate the molecules in alkynes and alkanes.

c

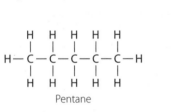

Pentane

2,2-Dimethylpropane

The two compounds will not have the same boiling point. Pentane will have a higher boiling point than 2,2-dimethylpropane. Both compounds have weak dispersion forces between their molecules; however, the straight chains of pentane can pack closer to form stronger overall dispersion forces than in the branched isomer, 2,2-dimethylpropane.

d 1 Place the mixture with some boiling chips in a 250 mL round bottom flask in a heating mantle.
 2 Connect a fractionating column to the round bottom flask.
 3 Connect a thermometer and Liebig condenser to the fractionating column and connect to a tap ready for distillation.
 4 Collect the distillate that distils first when the temperature is stable at 64°C. This is hex-1-ene.
 5 Collect the distillate that distils next when the temperature is stable at 69°C. This is hexane.
 6 The final distillate will be hex-1-yne, which will distil at 71°C.

2 1 Label the bottles 'X' and 'Y'.
 2 Pour about 25 mL of each liquid into a separating funnel.
 3 The immiscible liquids will form two layers.
 4 The bottom layer will be water as it is the denser of the two liquids. The top layer will be hept-1-yne.
 5 Match 'X' and 'Y' to the respective layers.

3 a

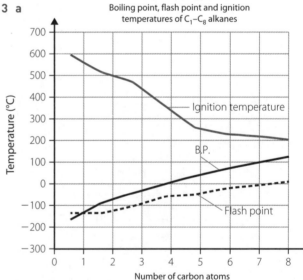

Boiling point, flash point and ignition temperatures of C_1–C_8 alkanes

b The flash point increases as boiling point increases, but ignition temperature decreases as boiling point increases. The longer the hydrocarbon chain, the stronger the dispersion forces between the non-polar molecules, so the less likely they are to be volatile, hence less likely to form a volatile combustible mixture. Ignition temperature, on the other hand, can be referred to as a measure of the activation energy of a combustion reaction.

c i The % composition of butane $= \dfrac{15}{200} \times 100 = 7.5\%$

  The mixture is combustible at 7.5% because it lies in the range given, 1.8–8.4%.

 ii Since the mixture is at 100°C, which is below the ignition temperature of 365°C, the mixture will not spontaneously ignite.

9780170449656

iii The flash point of butane is −60°C, which is below the temperature of the mixture at 100°C. Therefore, if a match is brought near it, the mixture will ignite.

WS 9.2 PAGE 108

1 a Catalytic cracking and thermal cracking

b Catalytic cracking uses catalysts such as zeolites at 500°C and moderately low pressures, whereas thermal cracking is carried out at higher temperatures in the range 450–750°C at high pressures of about 70 atmospheres.

2 a Liquid Y is water because hydrocarbons are non-polar and are not soluble in water. The shorter chain hydrocarbon is not soluble in water as it is collected by the downward displacement of water.

b $C_{12}H_{26}(l) \rightarrow C_2H_4(g) + C_{10}H_{22}(l)$

c X could to be a mixture of ethene and decane because two products are formed in the reaction. However, as the B.P. of decane is expected to be much greater than that of ethene, given its larger mass (B.P. is 174°C), it is unlikely this will be in a gaseous form so the student's prediction most likely is correct.

3 a Layer X – gas; Layer Y – oil; Layer Z – water. Gas is less dense than oil so is found as a top layer Water is the densest and thus is the bottom layer, while oil and water are also immiscible.

b Deepwater Horizon has been the largest oil spill in US history. This has resulted in loss of plant and animal life as well as loss of jobs in the fishing and tourism industries. Fish and crabs were subsequently found to have deformities. The environmental effect was severe with carcinogens entering phytoplankton, which is a food source for other aquatic organisms. This meant the entire food chain was affected. The significant economic and social damage from this event placed pressure on the welfare system and local communities as many livelihoods were affected.

4 a i Corrosive chemicals

ii Explosive

iii Flammable

iv Oxidiser

v Irritants, health hazard

vi Toxic or poisonous

vii Gases under pressure

viii Environmentally damaging

ix Health hazard

b i Use in a well-ventilated area. Keep away from naked flame. Do not pour down sink. Collect in bottle after use.

ii Use in a well-ventilated area.

iii Keep away from naked flame. May be poured down the sink.

iv Use in a well-ventilated area. Keep away from naked flame. Do not pour down sink. Collect in bottle after use.

Chapter 10: Products of reactions involving hydrocarbons

WS 10.1 PAGE 112

1

	Hydrocarbon		Reagent	Catalyst	Product(s) – names and formulae
a	Ethene	+	HCl(g)	$\longrightarrow$	Chloroethane
b	Propene	+	H_2(g)	Pt $\longrightarrow$	Propane
c	Hex-3-ene	+	Cl_2(g)	$\longrightarrow$	3,4-Dichlorohexane
d	Ethyne	+	$2F_2$(g)	$\longrightarrow$	1,2-Difluoroethene → 1,1,2,2-tetrafluoroethane
e	Ethene	+	H_2O(l)	dil. H_2SO_4 $\longrightarrow$	Ethanol
f	Ethyne	+	HCl(g)	$HgCl_2$ $\longrightarrow$	Chloroethene

2 a $CH_3CH_2CH_2CHCH_2(l) + Cl_2(g) \rightarrow CH_3CH_2CH_2CHClCH_2Cl(l)$

$CH_3CH_2CH_2CHCH_2(l) + H_2O(g) \rightarrow$
$CH_3CH_2CH_2CH_2CH_2OH(l)$

b X: pent-1-ene in excess

Y: pentan-1-ol + water (with dilute sulfuric acid)

Z: pent-1-ene + 1,2-dichloropentane

c The gas bubbled through conical flask A was steam with dilute sulfuric acid because it would form two immiscible layers of pent-1-ene (non-polar, less dense) in excess on the top and pentan-1-ol (polar, more dense) and water with dilute sulfuric acid in the bottom layer. The gas bubbled through conical flask B was chlorine because pent-1-ene and 1,2-dichloropentane would both be miscible.

3

		Major product	Minor product
a	Propene + HCl(g) →	2-Chloropropane	1-Chloropropane
b	But-1-ene + H_2O/dil. H_2SO_4	Butan-2-ol	Butan-1-ol

4 Hex-3-ene is a symmetrical alkene, so will only produce one product. When it is reacted with H_2O and dilute sulfuric acid, the product is hexan-3-ol.

EXTENSION

5 The two isomers are *cis*-but-2-ene and *trans*-but-2-ene, while the product is butane. There is only one product because there is free rotation around carbon–carbon single bonds.

cis-But-2-ene   *trans*-But-2-ene   Butane

WS 10.2 PAGE 115

1 a The reaction is classified as a substitution reaction because one of the hydrogen atoms attached to the carbon must be removed before the chlorine can attach to the carbon, Hence, a hydrogen is being substituted by a chlorine atom.

Equation: $CH_4(g) + Cl_2(g) \xrightarrow{UV} CH_3Cl(g) + HCl(g)$

b The UV light provides the minimum energy required for the reaction.

c The hydrocarbon shown, methane, is referred to as saturated because carbon can form four covalent bonds, and hence can bond with a maximum of four other atoms. In this case, the carbon has bonded with four hydrogen atoms.

2 a 1 $CH_4 + Cl_2 \rightarrow CH_3Cl + HCl$

2 $CH_3Cl + Cl_2 \rightarrow CH_2Cl_2 + HCl$

3 $CH_2Cl_2 + Cl_2 \rightarrow CHCl_3 + HCl$

4 $CHCl_3 + Cl_2 \rightarrow CCl_4 + HCl$

b $CH_4(g) + 4Cl_2(g) \rightarrow CCl_4(l) + 4HCl(g)$

3 a $n(\text{ethane}) = \dfrac{3.007}{(2 \times 12.01) + (6 \times 1.008)} = 0.1000 \, mol$

$n(\text{haloalkane product}) = n(\text{ethane}) = 0.1000 \, mol$

$MM(\text{halogenoalkane product}) = \dfrac{m}{n} = \dfrac{18.78}{0.100} = 187.8 \, g \, mol^{-1}$

Mass of two Br = $2 \times 79.90 = 159.8 \, g$

Then 187.8 – 159.8 = 28 g. This is equivalent to two carbon atoms and four hydrogen atoms. The molecular formula is $C_2H_4Br_2$.

b $C_2H_6 + 2Br_2 \rightarrow C_2H_4Br_2 + 2HBr$

4 a

1,1,1,2,2,3,3,3-Octabromopropane + hydrogen bromide

b

1-Chlorobutane + hydrogen chloride

252   ANSWERS   9780170449656

WS 10.3 PAGE 117

1 **a** The student was using bromine water in excess every time.

b 1 Label the bottles X and Y.

2 Label two small test tubes X and Y.

3 In a fume cupboard or in a well-ventilated area, in the absence of UV light, transfer 1 mL of each sample using a graduated pipette to the respective test tubes.

4 Add 2 or 3 drops of bromine water to each test tube and shake.

5 The test tube that has two colourless layers contains cyclohexene and the one that has a yellow orange colour in the top layer is cyclohexane.

c The hydrocarbons are non-polar, while the bromine water is polar; therefore, there are no intermolecular attractions between the layers. It was necessary to shake the test tube so the non-polar bromine could dissolve in the non-polar hydrocarbon by dispersion forces between molecules and react as appropriate.

d

Cyclohexene + Bromine water → 1,2-Dibromocyclohexane

e Cyclohexane and cyclohexene are liquids at room temperature and are therefore easier to handle than ethane and ethene, which are gases at room temperature. Cyclohexane and cyclohexene are also colourless so the colour change of bromine would be easier to observe. They are also relatively safe to handle.

f The waste would have been placed in specially labelled waste bottles for collection by a commercial company because toxic organic waste should not be poured down the sink.

2 **a** The bromine water would have changed colour from yellow-orange to colourless.

b The gas that did not react was a saturated hydrocarbon.

c The hydrocarbon gas is unsaturated as it reacts with bromine water in the absence of UV light.

$$n(\text{gas}) = \frac{V_{\text{gas}}}{V_{\text{molar}}} = \frac{0.558}{24.79} = 0.0255...$$

$$MM(\text{gas}) = \frac{m}{n} = \frac{0.631}{0.225} = 28.033...$$

The gas is ethene, C_2H_4.

d The gas reacted with the dilute sulfuric acid to produce ethanol, which is a liquid and soluble in water. Therefore, no gas was collected.

Chapter 11: Alcohols

WS 11.1 PAGE 120

1

	Structure	IUPAC name
a		Butan-2-ol
		Primary, secondary or tertiary alcohol
		Secondary
b		**IUPAC name**
		Methanol
		Primary, secondary or tertiary alcohol
		Primary
c		**IUPAC name**
		2-Methylpropan-2-ol
		Primary, secondary or tertiary alcohol
		Tertiary
d		**IUPAC name**
		Cyclohexanol
		Primary, secondary or tertiary alcohol
		Secondary
e		**IUPAC name**
		2-Methylpentan-3-ol
		Primary, secondary or tertiary alcohol
		Secondary

2 Hexane is an alkane so will be a non-polar solvent as it has only dispersion forces between molecules. Since hexane is non-polar and the hydrocarbon chain of the alcohol is also non-polar, there will be attraction between these two components, whereas the –OH part of the alcohol is polar so is attracted to polar solvents. Therefore, as the chain length of the alcohol increases, there would be greater interaction between the non-polar hydrocarbon end of the alcohol and the hexane, and hence, there would be an increase in solubility of the alcohol in hexane.

3 a

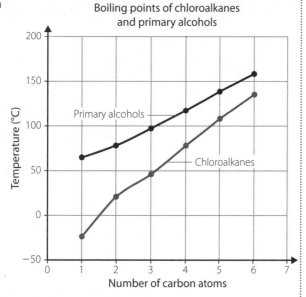

Boiling points of chloroalkanes and primary alcohols

(Graph: y-axis "Temperature (°C)" from −50 to 200; x-axis "Number of carbon atoms" 0 to 7. Two curves labelled "Primary alcohols" and "Chloroalkanes".)

b The boiling points for both the chloroalkanes and primary alcohols increase as the number of carbon atoms increases. The boiling points of the primary alcohols are consistently higher than the boiling points of the chloroalkanes due to the strong hydrogen bonds between the –OH groups on different alcohol molecules; therefore, more energy is required to overcome these stronger intermolecular forces. The chloroalkanes have dipole–dipole interactions between molecules due to the polar nature of the carbon–chlorine bond. Because these dipole–dipole bonds are weaker than hydrogen bonds, less energy is required to overcome these forces and separate the molecules.

As chain length increases, the difference in boiling points between chloroalkanes and primary alcohols decreases because hydrogen bonds and dipole–dipole forces have a smaller effect on the overall intermolecular forces than dispersion forces do.

4 The three isomers all have covalent bonds as their intramolecular forces. They will not have the same boiling point because the strength of intermolecular forces determines the boiling point of substances. All three molecules have dispersion forces as well as hydrogen bonds between their molecules. The position of the hydroxyl group and the degree of branching in the hydrocarbon chain need to be considered as the hydroxyl group forms the strongest intermolecular force with neighbouring molecules. In pentan-1-ol, the hydroxyl group is most easily accessible. Therefore, the hydroxyl groups can form strong hydrogen bonds with neighbouring molecules and more energy would be required to separate the molecules, thus, would have the highest boiling point. In 2-methylbutan-2-ol, the OH group is not as accessible as the OH group on on pentan-2-ol and there is less interaction between the molecules because of the branching in the hydrocarbon chain, so the boiling point of 2-methytlbutran-2-ol is lower than that of pentan-2-ol.

5 a Alcohol group in both. Testosterone has a ketone and an alkene group, while oestradiol has a benzene ring.

b Alcohol groups are secondary in both compounds.

c Melting point is determined by the strength of intermolecular forces. Oestradiol has a higher melting point than testosterone because in addition to dispersion forces it has the strongest intermolecular force, hydrogen bonds occurring between two sections of neighbouring molecules. Testosterone has hydrogen bonds occurring between one section and weaker dipole–dipole forces between neighbouring carbonyl groups. Both oestradiol and testosterone experience dispersion forces, with testosterone experiencing slightly greater dispersion forces because it has

a higher molar mass. However, since dispersion forces are the weakest of intermolecular forces, their effect is not as significant as hydrogen bonds or dipole–dipole forces.

d Ethanol is a small molecule that has both polar and non-polar parts, so it can dissolve both polar and non-polar solutes. Both oestradiol and testosterone have greater non-polar sections in their molecules compared to the polar hydroxyl part. Since the non-polar part of ethanol can form dispersion forces with non-polar sections of solutes and the hydroxyl part can form hydrogen bonds with the hydroxyl groups, it can dissolve both these hormones.

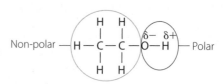

WS 11.2 PAGE 124

1 a $CH_3CH_2CH_2OH(l) + 4.5O_2(g) \rightarrow 3CO_2(g) + 4H_2O(l)$

b i Ali

$m(\text{water}) = 125\,g$

$\Delta T(\text{water}) = 34 - 22 = 12°C$

$q(\text{absorbed by water}) = mc\Delta T = 0.125 \times 4.18 \times 10^3 \times 12\,J$

$q\,(\text{released by propan-1-ol}) = q(\text{absorbed by water})$

$m(\text{propan-1-ol}) \text{ used} = 265.243 - 264.884 = 0.359\,g$

$n(\text{propan-1-ol}) \text{ used} = \dfrac{m}{MM} = \dfrac{0.359}{(3 \times 12.01) + (8 \times 1.008) + 16.00}$
$= 0.005\,97$

$\Delta H = \dfrac{q}{n}$

$\Delta H = \dfrac{0.125 \times 4.18 \times 10^3 \times 12.0}{0.005\,97} = 1.050\,25.... \times 10^6\,J\,mol$

$= 1050\,kJ\,mol$

ii Sam

$m(\text{water}) = 125\,g$

$\Delta T = 34 - 22 = 12°C$

$m(\text{propan-1-ol}) \text{ used} = 268.357 - 268.148 = 0.209\,g$

$n(\text{propan-1-ol}) \text{ used} = \dfrac{m}{MM} = \dfrac{0.209}{(3 \times 12.01) + (8 \times 1.008) + 16.00}$
$= 0.003\,48$

$\Delta H = \dfrac{q}{n}$

$q = mc\Delta T = 0.125 \times 4.18 \times 10^3 \times 12.0\,J$

$\Delta H = \dfrac{0.125 \times 4.18 \times 10^3 \times 12.0}{0.003\,48} = 1.801\,72... \times 10^6\,J\,mol$

$= 1802\,kJ\,mol$

c i Ali's % error =

$\dfrac{\text{theoretical} - \text{experimental}}{\text{theoretical}} \times 100 = \dfrac{2021 - 1050}{2021} \times 100 = 48.05\%$

Sam's % error =

$\dfrac{\text{theoretical} - \text{experimental}}{\text{theoretical}} \times 100 = \dfrac{2021 - 1802}{2021} \times 100 = 10.84\%$

ii Sam's percentage error of 10.84% is less than Ali's 48.05% because Sam's experimental set-up minimised heat loss to the environment. Sam's results are closer to the theoretical value because Sam used a copper calorimeter, which transfers heat well, and being enclosed with a draught shield and insulating card, it did not lose as much heat. Ali's set-up resulted in heat being lost to the tripod, gauze mat and glass beaker, which would not have transferred heat to the water as efficiently because glass is an insulator. The top of the beaker was open, and no

9780170449656

draught shield was used to prevent further heat loss. Both experiments were valid because the set-up allowed for the molar enthalpy of combustion of propan-1-ol to be calculated. Neither experiment was reliable as they were not repeated. Neither experiment was accurate due to the relatively high percentage error calculated. Furthermore, since the bottom of Ali's beaker was covered in more soot than Sam's copper calorimeter, his alkanol underwent a greater degree of incomplete combustion.

2 The presence of soot indicates incomplete combustion. Methanol requires the least amount of oxygen per mole, at 1.5 moles for complete combustion, while pentan-1-ol requires the greatest amount of oxygen at 7.5 moles as shown by the equations below. Therefore, methanol is most likely to undergo complete combustion compared to ethanol and pentan-1-ol and produce the least soot, while pentan-1-ol is most likely to undergo incomplete combustion.

$$CH_3OH(l) + 1.5O_2(g) \rightarrow CO_2(g) + 2H_2O(l)$$
$$C_2H_5OH(l) + 3O_2(g) \rightarrow 2CO_2(g) + 3H_2O(l)$$
$$C_5H_{11}OH(l) + 7.5O_2(g) \rightarrow 5CO_2(g) + 6H_2O(l)$$

Therefore, the tin with the least soot at the base, tin Y, matches methanol. The tin with moderate soot at the base, tin X, matches ethanol while the tin with the most soot, tin Z, matches pentan-1-ol.

WS 11.3 PAGE 127

1 a $C_2H_5OH(l) + 3O_2(g) \rightarrow 2CO_2(g) + 3H_2O(g)$

b $C_4H_9OH(l) + 3.5O_2(g) \rightarrow C(s) + 3CO(g) + 5H_2O(g)$

c $CH_3CHOHCH_3(l) \xrightarrow{\text{conc. }H_2SO_4} CH_3CHCH_2(g) + H_2O(l)$
propene

d $CH_3OH(l) + HCl(g) \rightarrow CH_3Cl(l) + H_2O(l)$
chloromethane

e
$CH_3 CHOH CH_2 CH_2 CH_2CH_3 (l) \xrightarrow{Cr_2O_7{}^{2-}/H^+} CH_3 CO CH_2 CH_2 CH_2CH_3 (l)$
Hexan-2-one

f No visible reaction as 2-methylpropan-2-ol is a tertiary alcohol, which cannot be oxidised under these conditions.

2 Addition of bromine water

$CH_3CH_2CHCH_2(g) + Br_2(aq) \rightarrow CH_3CH_2CHBrCH_2Br(g)$
1,2-dibromobutane

$CH_3CHCHCH_3(g) + Br_2(aq) \rightarrow CH_3CHBrCHBrCH_3(g)$
2,3-Dibromobutane

3 $C_xH_yO + O_2(g) \rightarrow xCO_2(g) + \frac{y}{2}H_2O(l)$

$n(CO_2) = \frac{m}{MM} = \frac{0.5501}{44.01} = 0.012\,499... \text{ mol}$

$n(C) = n(CO_2) = 0.012\,499... \text{ mol}$

$m(C) = n \times M = 0.012\,499... \times 12.01 = 0.1501 \text{ g}$

$n(H_2O) = \frac{m}{MM} = \frac{0.2702}{18.016} = 0.014\,997... \text{ mol}$

$n(H) = 2 \times n(H_2O) = 2 \times 0.014\,997... = 0.029\,99... \text{ mol}$

$m(H) = n \times MM = 0.030\,2355... = 0.030\,24... \text{ g}$

$m(O) = 0.2203 - [m(C) + m(H)]$
$\quad\quad = 0.2203 - [0.1501 + 0.030\,24]$
$\quad\quad = 0.039\,96 \text{ g}$

$n(O) = \frac{m}{MM} = \frac{0.039\,96}{16.00} = 0.002\,4975 \text{ mol}$

	C	H	O
n	0.012 499...	0.029 99...	0.002 4975...
Divide by 0.002 4975	5.0046	12.008	1
Whole number ratio	5	12	1

The primary alcohol has 5 carbon atoms, 12 hydrogen atoms and one oxygen atom, so it must be pentan-1-ol, $C_5H_{11}OH$.

4 Addition reaction

Substitution reaction

5 a 1 Label the bottles 'X', 'Y' and 'Z'.

2 Label three test tubes 'X', 'Y' and 'Z' and add 1 mL of water to each test tube.

3 Add 4–5 drops of the liquids X, Y and Z to the appropriate test tubes.

4 The liquid that is immiscible is octan-1-ol.

5 Label two more test tubes and place 1 mL of acidified potassium dichromate in each test tube.

6 Add 2 mL of the remaining two liquids to the corresponding test tubes and heat the test tubes.

7 The test tube that contains the mixture that remains orange colour contains 2-methylpropan-2-ol.

8 The test tube in which the mixture has changed from orange to green contains ethanol.

b

Observation	Explanation
Octan-1-ol is immiscible in water.	The non-polar hydrocarbon part of octan-1-ol is not able to form intermolecular forces with water and while the hydroxyl group can form hydrogen bonds with water, it is not strong enough to make octan-1-ol soluble in water.
2-Methylpropan-2-ol does not react with the acidified potassium dichromate, therefore, the orange colour remains.	2-Methylpropan-2-ol is a tertiary alcohol and cannot be oxidised by acidified potassium dichromate.
Ethanol reacts with acidified potassium dichromate, changing it from orange to green.	Ethanol is a primary alcohol, which can be oxidised by acidified potassium dichromate. Therefore, the orange dichromate ions area reduced to green Cr^{3+} ions.

WS 11.4 PAGE 130

1 a Aqueous solution of glucose and alcohol-tolerant yeast

b $C_6H_{12}O_6(aq) \rightarrow 2CH_3CH_2OH(aq) + 2CO_2(g)$

c Aqueous solution of glucose (10% w/v), alcohol-tolerant yeast, temperature: about 37°C, anaerobic conditions (closed vessels; absence of air). *Optional: Nutrients for yeast (Pasteur's salt)*

d **i** Fermentation occurred in flask X as the reaction mixture was under anaerobic conditions and the mass of the flask decreased due to loss of carbon dioxide. Fermentation was complete by day 3 in flask X as can be seen in the constant mass. There was no fermentation in flask Y as it was stoppered with cotton wool, which enabled air to pass through it; hence, it was not under anaerobic conditions. The slight variation in mass was possibly due to evaporation. An improvement would have been to include a control to see if any mass was lost due to evaporation.

   ii Limewater was used to confirm the production of carbon dioxide gas because limewater would turn milky in the presence of carbon dioxide then colourless again.

   iii $m(CO_2) = 250.12 - 243.16\,g = 6.96\,g$

   $n(CO_2) = \dfrac{6.96}{12.01 + 2 \times 16.00} = 0.1581...\,mol = n(\text{ethanol})$

   $n(C_6H_{12}O_6)\text{fermented} = \dfrac{0.1581...}{2} = 0.079\,07...\,mol$

   $MM(\text{glucose, } C_6H_{12}O_6) = (6 \times 12.01) + (12 \times 1.008) + (6 \times 16.00) = 180.156\,g\,mol^{-1}$

   $m(C_6H_{12}O_6)\text{fermented} = 0.079\,07.... \times MM(\text{glucose}) = 14.245...\,g$

   $\%(\text{glucose})\text{fermented} = \dfrac{14.245}{97.57} \times 100 = 14.599...\%$
   $= 14.60\,\%$ (to 4 sig fig)

   The percentage of glucose that was fermented was 14.60%.

   e Yeast die when the ethanol concentration reaches 15%; hence, without yeast, the fermentation process cannot continue.

   f Filter the fermentation mixture, then fractionally distil the filtrate. The fraction that distils at 78°C should be collected because it will be the ethanol distillate.

2 Addition reaction:
$CH_2CH_2(g) + H_2O(l) \xrightarrow{\text{dil. }H_2SO_4} CH_3CH_2OH(aq)$

Substitution reaction: $CH_3CH_2Cl(g) + NaOH(aq) \rightarrow CH_3CH_2OH(aq) + NaCl(aq)$

3 Since the alcohol could not be oxidised, it must be a tertiary alcohol.

$(CH_3)_3CCl + H_2O(l) \rightarrow (CH_3)_3COH + HCl(aq)$

2-chloro-2-methylpropane   2-methylpropan-2-ol

$(CH_3)_3CCl + NaOH(aq) \rightarrow (CH_3)_3COH + NaCl(aq)$

2-chloro-2-methylpropane   2-methylpropan-2-ol

WS 11.5 PAGE 132

1 **a** Heat the reaction mixture so it is above the boiling point of ethanal; therefore, as soon as ethanal is produced, it boils and passes through the condenser to be collected in the second flask that is cooled in the ice water.

   b The ethanol cannot be oxidised to ethanoic acid using the set-up shown because the intermediary product, ethanal, is removed as soon as it forms.

   c Ethanol is volatile and flammable; therefore, a naked flame, such as that in a Bunsen burner, cannot be used.

   d Oxidation half-equation: $(CH_3CH_2OH(l) \rightarrow CH_3CHO(l) + 2H^+(aq) + 2e^-) \times 3$
   Reduction half-equation: $Cr_2O_7^{2-}(aq) + 14H^+(aq) + 6e^- \rightarrow 2Cr^{3+}(aq) + 7H_2O$

NET equation: $3CH_3CH_2OH(l) + Cr_2O_7^{2-}(aq) + 8H^+(aq) \rightarrow 3CH_3CHO(l) + 2Cr^{3+}(aq) + 7H_2O(l)$

   e The orange colour of the potassium dichromate would change to green as Cr^{3+} ions form.

   f Oxidant: acidified dichromate ions, $H^+/Cr_2O_7^{2-}$
   Reductant: ethanol, CH_3CH_2OH

2 **a** X: Reflux condenser
   Y: Water out
   Z: Water in

   b Propan-1-ol is first oxidised to propanal, which has a lower boiling point than the reaction mixture. However, it cannot escape because it condenses back in the reaction mixture due to the reflux condenser cooling down the vapours. The propanal can then be oxidised to propanoic acid by the acidified dichromate ions.

   c Oxidation half-equation: $(CH_3CH_2CH_2OH(l) + H_2O(l) \rightarrow CH_3CH_2COOH(l) + 4H^+(aq) + 4e^-) \times 3$
   Reduction half-equation: $(Cr_2O_7^{2-}(aq) + 14H^+(aq) + 6e^- \rightarrow 2Cr^{3+}(aq) + 7H_2O) \times 2$
   NET equation: $3CH_3CH_2CH_2OH(l) + 2Cr_2O_7^{2-}(aq) + 16H^+(aq) \rightarrow 3CH_3CH_2COOH(l) + 4Cr^{3+}(aq) + 11H_2O(l)$

   d Porcelain/glass boiling chips should be added to the reaction flask to promote even boiling.

3 **a** The purple colour of $KMnO_4$ changed to very pale pink/almost colourless as it was reduced.

   b The student's conclusion is not valid because she did not check the boiling point of butan-2-one in the literature, which is the product of oxidation of butan-2-ol.
   The oxidation half-equation is $(CH_3CH(OH)CH_2CH_3(l) \rightarrow CH_3COCH_2CH_3(l) + 2H^+ + 2e^-) \times 5$
   NET equation: $5CH_3CH(OH)CH_2CH_3(l) + 2MnO_4^-(aq) + 6H^+(aq) \rightarrow 5CH_3COCH_2CH_3(l) + 2Mn^{2+}(aq) + 8H_2O(l)$
   The student can further check her conclusion by refluxing the product using more acidified potassium permanganate and separating and testing the product with solid Na_2CO_3. If bubbling occurs, this would confirm the presence of butanoic acid, which can be produced by further oxidation of butanal.

WS 11.6 PAGE 135

1 **a** A fuel is a substance that releases energy when it undergoes combustion.

   b Coal and natural gas are two examples of fossil fuels.
   Current concern with their use is that they are non-renewable, so fossil fuels will eventually run out and they produce environmental pollutants such as carbon dioxide that give rise to the enhanced greenhouse effect.

   c Biofuels are produced from organic sources like crops or sugars. Renewable sources such as algae, animal wastes and sugar cane can be used. The two main biofuels used in Australia are biodiesel and ethanol.

2 Diesel is a mixture of long chain hydrocarbons with 8–21 carbons. Biodiesel is similar but contains an ester group and is formed from the breakdown of triglycerides.

3 **a** Transesterification is the name of the reaction.

   b X: Saturated fatty acids
   Y: Glycerol
   Z: Triglyceride

Reactant 1	**Reactant 2**	**Conditions**
(structure of triglyceride)	$3CH_3OH$	NaOH/KOH

Product 1 (biodiesel)

(structure, with 3 and H_3C ... $O-CH_3$)

c   Product 2

$$HO-CH_2$$
$$HO-CH$$
$$HO-CH_2$$

---

d The use of NaOH and KOH is economical because it has a 98% conversion yield and requires low temperatures. However, it produces high temperatures, which pose a safety risk, and forms soap as a by-product, reducing the yield of the biodiesel. It is also expensive to purify the glycerol from the base. Lipase is expensive to produce but a higher yield of biodiesel is produced because soap is not produced as a by-product.

4 a Bioethanol: $C_2H_5OH(l) + 3O_2(g) \rightarrow 2CO_2(g) + 3H_2O(l)$

m(bioethanol) combusted $= 0.78 \times 1000 \times 60 = 46\,800\,g$

n(bioethanol) $= \dfrac{46\,800}{46.068} = 1015.88\ldots\,mol$

$n(CO_2) = 2 \times n$(bioethanol) $= 2 \times 1015.88\ldots = 2031.779\ldots\,mol$

$V(CO_2) = n \times 24.79\,L = 50\,367.80\ldots\,L = 5.0 \times 10^4\,L$ (to 2 sig fig)

Octane: $C_8H_{18}(l) + 12.5O_2(g) \rightarrow 8CO_2(g) + 9H_2O(l)$

m(octane) combusted $= 0.69 \times 1000 \times 60 = 41\,400\,g$

n(octane) $= \dfrac{41400}{114.224} = 362.445\ldots\,mol$

$n(CO_2) = 8 \times n$(octane) $= 8 \times 362.445\ldots = 2899.565\ldots\,mol$

$V(CO_2) = n \times 24.79\,L = 71\,880.235\ldots\,L = 7.2 \times 10^4\,L$ (to 2 sig fig)

The volume of carbon dioxide produced by bioethanol is about $\left(\dfrac{7.2-5.0}{7.2} \times 100\right) = 31\%$, less than that produced by octane.

b Octane releases $48\,kJ\,g^{-1}$, while bioethanol releases $29.6\,kJ\,g^{-1}$.

m(bioethanol) $= \dfrac{48}{29.6} = 1.621\ldots\,g$

V(bioethanol) $= \dfrac{mass}{density} = \dfrac{1.621\ldots\,g}{0.78\,g\,L^{-1}} = 2.079\ldots\,L = 2.1\,L$ (to 2 sig fig)

Chapter 12: Reactions of organic acids and bases

WS 12.1 PAGE 138

1

Label	Name	Structure
1	But-1-ene	(structure)
2	1-Chlorobutane	(structure)
3	Butanamine	(structure)
4	Butanol	(structure)
5	Butanoic acid	(structure)

Label	Name	Structure
6	Butan-2-ol	
7	Butan-1-ol	
Reagent X	Conc. H_2SO_4	

2 a

Pentanamine

N-Methylpropanamine

N,N-Dimethylpropanamine

b The lowest boiling point is that of *N,N*-dimethylpropanamine as it cannot form hydrogen bonds between its molecules because there is no H attached to a N, which is necessary for H bonding to occur. The next highest boiling point is that of *N*-methylpropanamine as it can form hydrogen bonds but this part of the molecule is not as accessible. Pentanamine has the highest boiling point because it can form hydrogen bonds between its molecules most easily; therefore, most energy is required to separate the molecules.

3 a Hydroxyl group

Carboxylic acid

b Amoxicillin is soluble in water as the hydroxyl group, amine group and carboxylic acids can form hydrogen bonds with water, while the carbonyl group of the amide and the β-lactam ring can form dipole–dipole interactions with water, overcoming the large molecular mass and large non-polar sections of the molecule that would otherwise make the molecule insoluble.

4 a From the graph:
c(propanone) $= 3.8\,\text{mM}$
c(stearic acid) $= 1.2\,\text{mM}$

In the 10 mL plasma sample:
n(propanone) $= cV = 3.8 \times 10^{-3} \times 0.010 = 3.8 \times 10^{-5}\,\text{mol}$
m(propanone) $= n \times MM = 3.8 \times 10^{-5} \times ((3 \times 12.01) + 16.00$
$\qquad + (6 \times 1.008))$
$\qquad = 2.206... \times 10^{-3}\,\text{g}$
$\qquad = 2\,\text{mg (to 1 sig fig)}$
n(stearic) $= cV = 1.2 \times 10^{-3} \times 0.010 = 1.2 \times 10^{-5}\,\text{mol}$
m(stearic acid) $= n \times MM = 1.2 \times 10^{-5} \times ((18 \times 12.01)$
$\qquad + (2 \times 16.00) + (36 \times 1.008))$
$\qquad = 3.413... \times 10^{-3}\,\text{g}$
$\qquad = 3\,\text{mg (to 1 sig fig)}$

On the fifth day of starvation, the mass of propanone would be 2 mg and the mass of stearic acid would be 3 mg.

b Propanone is more soluble in water than stearic acid because propanone is a small molecule with a polar carbonyl group that can form dipole–dipole interactions with polar water. Stearic acid has a long non-polar hydrocarbon section and only a smaller polar carboxyl group; hence, the predominant non-polar section of the molecule makes it less soluble in water.

5 a The carbonyl group requires a carbon atom on either side to be a ketone; hence, minimum of three carbon atoms are required.

b

c Boiling point is determined by the strength of the intermolecular forces. The greater the intermolecular force, the higher the boiling point as more energy is required to overcome the intermolecular forces. As the number of carbon atoms increases, so does the boiling point of every group because the strength of the dispersion forces increases. Aldehydes and ketones (carbonyl group is more polarised in ketones) have similar boiling points, which are the lowest for the functional groups shown, as they both contain the polar carbonyl group with dipole–dipole forces whereas carboxylic acids and amides have hydrogen bonds in addition to dipole–dipole forces. Amides have three possible sections for hydrogen bonds as part of the –NH_2 while carboxylic acids have two possible sections for hydrogen bonds as part of the –OH group. Hence, the amides have stronger intermolecular forces than carboxylic acids and therefore more energy is required to overcome these intermolecular forces, causing amides to have higher boiling points than carboxylic acids.

1 a

Butan-1-ol   Ethanoic acid   Butyl ethanoate

b i 1 Reflux condenser
  2 Cold water in from tap
  3 Round-bottom flask
  4 Reaction mixture
  5 Boiling chips (porcelain)
  6 Heating mantle
  7 Water out to sink

  ii Refluxing is the name of the process used to make the ester butyl ethanoate. Esterification is slow at room temperature, so the reactants need to be heated for an extended amount of time and a catalyst added to speed up the rate of reaction. However, since the reactants are volatile and flammable, a reflux condenser is used so that the reactants can condense back into the round-bottom flask to continue reacting to form an ester. A heating mantle is used to avoid naked flames. Concentrated sulfuric acid is used as a catalyst because it provides an alternative pathway for the reaction to occur that has a lower activation energy. It also acts as a dehydrating agent, absorbing the water formed to drive the reaction to the right by Le Chatelier's principle, increasing the yield of the ester.

c Butan-1-ol, ethanoic acid, sulfuric acid, water, butyl ethanoate

~continue in right column

d i The equipment shown is a separating funnel.
  ii Two immiscible layers form because the ester is not soluble in water as it is mainly non-polar. However, the water-soluble components, ethanoic acid and butan-1-ol, dissolve in the water layer by forming hydrogen bonds. The ethanoic acid also forms dipole–dipole interactions with water.
    X: butyl ethanoate
    Y: butan-1-ol, ethanoic acid, sulfuric acid, water

e The aqueous sodium carbonate reacts with any remaining acid and the bubbling is due to the formation of carbon dioxide gas in the reaction:
$$2H^+(aq) + CO_3^{2-}(aq) \rightarrow CO_2(g) + H_2O(l)$$

f Fractional distillation.

2 1 Label the beakers A, B and C.
  2 Pour 2–3 drops from each beaker into three depression tiles labelled A, B and C and test with litmus paper.
  3 The liquid in which litmus paper turns red is methanoic acid.
  4 In two test tubes, labelled appropriately, add 1 mL of distilled water.
  5 Add 1 mL of the remaining unidentified chemicals to each test tube.
  6 The test tube that has one homogeneous layer contains ethanol, while the test tube with two immiscible layers contains ethyl methanoate.

3

	Alkanol	Carboxylic acid	Ester	Water
a	Methanol	Hexanoic acid	Methyl hexanoate	
b	Propan-1-ol	Butanoic acid	Propyl butanoate	
c	Ethanol	Pentanoic acid	Ethyl pentanoate	

1 Saponification is the conversion, under alkaline conditions, of fats or oils into glycerol and salts of fatty acids.

2

Triglyceride   Glycerol   Soap

3 Soap becomes protonated in acidic solutions forming a molecule that no longer has a charged head and is not soluble in water and can no longer act as soap.

$$RCOO^-Na^+ + H^+ \rightarrow RCOOH + Na^+$$

In hard water, the calcium and magnesium salts of soap are not soluble; hence, a precipitate forms, which can no longer act as soap.

$$2RCOO^-Na^+ + Ca^{2+} \rightarrow Ca(RCOO)_2(s) + 2Na^+$$

4 The non-polar ends of the soap ions bond to the surface of the grease via dispersion forces.

The ionic ends bond to water molecules via H-bonding/ion–dipole interaction.

The grease on clothing or on skin is loosened by agitation or rubbing.

The non-polar ends of the soap ions surround more and more of the grease droplet as it is loosened from the surface by further agitation.

The result is a micelle, with a spherical shape – the grease droplet is surrounded by the non-polar ends of many soap ions, with the ionic ends bonding to water.

Thus, the grease is dispersed throughout the water in the form of an emulsion – the emulsion is the soap, water and grease mixture.

5 a i Cationic
   ii Anionic
   iii Non-ionic

  b i Anionic detergents can be used as general laundry detergents as they are effective in hard water and produce sufficient froth.

ii Cationic detergents can be used as fabric or hair conditioners as they have a positive head and are attracted to negative surface charges on fibres in clothing and hair, thereby forming a smooth layer on top. Cationic detergents can also be used as disinfectants as in nappy wash detergents.

iii Non-ionic detergents can be used as detergents in front-loading washing machines as they are low froth.

6 1 Label the beakers A and B.
   2 Label two test tubes A1 and A2 and two test tubes B1 and B2.
   3 Using a 10 mL measuring cylinder, pour 2 mL of each liquid into the appropriate test tube.
   4 Add 4 mL of distilled water to A1 and B1. Stopper and shake the test tubes. Record the height of the bubbles.
   5 Using a 10 mL measuring cylinder, pour 4 mL of 1 mol L^{-1} calcium nitrate or 1 mol L^{-1} magnesium nitrate into A2 and B2. Stopper and shake the test tubes. Record the height of the bubbles.
   6 The test tube with least bubbles when 1 mol L^{-1} calcium nitrate or 1 mol L^{-1} magnesium nitrate is added is the soap, while the other one is the cationic detergent.

   Justification

   The soap would form a precipitate with the calcium/magnesium ions and would not froth as much as in the control test tube in which the soap was shaken with the distilled water. The amount of froth produced by the cationic detergent would be about the same in both distilled water and in the 1 mol L^{-1} calcium nitrate or 1 mol L^{-1} magnesium nitrate because the cationic head does not react with the positive calcium or magnesium ions.

Chapter 13: Polymers

WS 13.1 PAGE 148

1 A monomer needs to have a double bond in order to form an addition polymer.

2

Monomer name		Monomer structure	Polymer repeat unit structure	Polymer name	Use
Common	Systematic				
Ethene Ethylene	**Ethene**			**Polyethene**	**LDPE – cling wrap**
					HDPE – kitchen utensils
Vinyl chloride	**Chloroethene**			**Poly(vinyl chloride)**	**Electrical insulation**
Styrene	Ethenylbenzene			**Polystyrene**	**Foam insulation/ tool handles**
Tetrafluoroethene	Tetrafluoroethene			**Polytetrafluoroethene**	**Non-stick surfaces on cookware**

3 $MM(\text{TFE}) = 100.02\,\text{g mol}^{-1}$

$n(\text{TFE})$ in PTFE $= \dfrac{MM(\text{PTFE})}{MM(\text{TFE})} = \dfrac{5.5511 \times 10^4}{100.02} = 555\,\text{mol}$

$n(\text{F}) = 4 \times n(\text{TFE}) = 4 \times 555 = 2220\,\text{mol}$

The number of fluorine atoms in an average molecule of PTFE is 2220.

4 a X is HDPE and Y is LDPE. Structure X has polymer chains in a more ordered crystalline structure, while structure Y has more branching of the polymer chains and it is a more random arrangement of the chains in an amorphous/non-crystalline structure.

b Structure X has the polymer chains closely packed in an ordered structure so it is not as flexible as structure Y that has its chains branched and in a more random structure. Structure X will also have a higher melting point than structure Y due to stronger dispersions forces between the closely ordered chains compared to structure Y.

c Structure Y, LDPE, is produced through initiation, propagation and termination. The final step, termination, is random; hence, a variety of molecular masses are observed in the sample.

5

	Monomer	Polymer repeat unit
a		
b		
c	CH₂=CH—O—CO—CH₃	

6

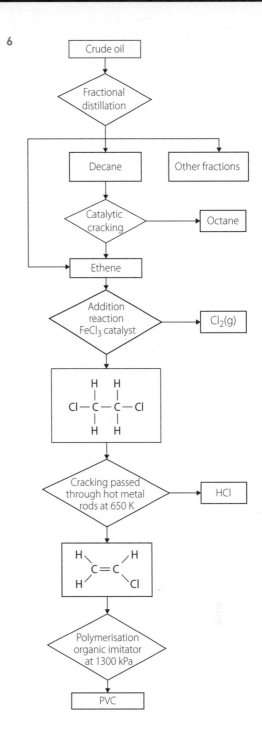

WS 13.2 PAGE 151

1 a

Amide link

b

Ester link

2

	Monomer(s)	Polymer repeat unit
a	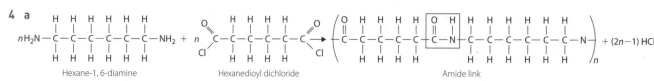	
b		
c		
d		

3 a

Peptide link

b MM(glycine) $= (2 \times 12.01) + (5 \times 1.008) + (2 \times 16.00) + 14.01 = 75.07\,\text{g mol}^{-1}$

MM(glycine) $= 75.07 \times 600 = 45\,042\,\text{g mol}^{-1}$

MM(alanine) $= (3 \times 12.01) + (7 \times 1.008) + (2 \times 16.00) + 14.01 = 89.096\,\text{g mol}^{-1}$

MM(alanine) $= 89.096 \times 600 = 53\,457.6\,\text{g mol}^{-1}$

$MM(H_2O) = (2 \times 1.008) + 16.00 = 18.016\,\text{g mol}^{-1}$

$MM(H_2O)$ formed $= 1199 \times 18.016 = 21\,601.184\,\text{g mol}^{-1}$

Mass of polypeptide formed $= (45\,042 + 53\,457.6) - 21\,601.184\,\text{g}$

Mass of polypeptide formed $= 76\,898.416\,\text{g} = 7.69 \times 10^4\,\text{g}$ (to 3 sig fig)

The average mass of the polypeptide formed when 600 moles each of glycine and alanine polymerise is $7.69 \times 10^4\,\text{g}$.

~continue in right column ▲

4 a

Hexane-1, 6-diamine   Hexanedioyl dichloride   Amide link

b MM(HCl) $= 1.008 + 35.45 = 36.458$

n(HCl) $= \dfrac{18\,229}{36.458} = 500$

If $2n - 1 = 500$, then $n = 250.5$

MM(hexane-1,6-diamine) $= (6 \times 12.01) + (16 \times 1.008) + (2 \times 14.01) = 116.208$

m(hexane-1,6-diamine) $= 250.5 \times 116.208 = 29\,110.104$

MM(hexanedioyl chloride) $= (6 \times 12.01) + (8 \times 1.008) + (2 \times 16.00) + (2 \times 35.45) = 183.024$

m(hexanedioyl chloride) $= 250.5 \times 183.024 = 45\,847.512$

Mass of nylon 6,6 formed $= (29\,110.104 + 45\,847.512) - 18\,229\,\text{g} = 56\,728.616\,\text{g}$

Mass of nylon 6,6 formed $= 5.673 \times 10^4\,\text{g}$ (to 4 sig fig because MM in periodic table are to 4 sig fig)

The mass of the polymer formed when $18\,229\,\text{g}$ of hydrogen chloride forms is $5.673 \times 10^4\,\text{g}$.

5 Synthetic polymers are made in the laboratory, while naturally occurring polymers occur in nature. Examples of synthetic condensation polymers are nylon 6,6 and polyethylene terephthalate, both of which are used to make fibres for textiles. Proteins and cellulose are examples of naturally occurring polymers. Proteins are essential in chemical processes that take place in the body, including transporting oxygen, while cellulose is necessary for supporting cell walls in plants. Synthetic polymers are not biodegradable and contribute to landfill, while naturally occurring polymers are biodegradable.

....

MODULE SEVEN: CHECKING UNDERSTANDING PAGE 154

1 C	**2** B	**3** D	**4** D	**5** C
6 C	**7** D	**8** C	**9** A	

10 a V(ethanol)used $= 341.26 - 335.65\,\text{mL} = 5.61\,\text{mL}$

m(ethanol)used $= V \times \text{density} = 5.61 \times 0.79\,\text{g} = 4.4319\,\text{g}$

n(ethanol) used $= \dfrac{m}{MM} = \dfrac{4.4319}{46.068} = 0.096\,203\ldots$

$\Delta H = \dfrac{0.255 \times 4.18 \times 10^3 \times 21.23}{0.096\,203}$

$= 235\,222\ldots\,\text{J mol}^{-1}$

$= -235.2\,\text{kJ mol}^{-1}$

b Percentage error $= \dfrac{|1367 - 235.2|}{1367} \times 100 = 82.79\%$

The student's value of 235.2 kJ mol^{-1} for the molar enthalpy of combustion of ethanol is lower than the theoretical value of 1367 kJ mol^{-1}. The lower experimental value is due to heat loss to the environment as no draught shield was used and the copper can was not covered. There may have also been incomplete combustion of ethanol. Improvements could include using a draught shield, covering and insulating the copper can and ensuring a sufficient supply of oxygen gas to ensure complete combustion. Also, the heat absorbed by the copper can could have been included in the calculation.

11 **a** $CH_3CH(OH)CH_3 \rightarrow CH_3COCH_3 + 2H^+ + 2e^-$

b $MnO_4^-(aq) + 8H^+(aq) + 5e^- \rightarrow Mn^{2+}(aq) + 4H_2O(l)$

c $5CH_3CH(OH)CH_3 + 2MnO_4^-(aq) + 6H^+(aq) \rightarrow 5CH_3COCH_3 + 2Mn^{2+}(aq) + 8H_2O(l)$

d The purple colour solution changes to pale pink or almost colourless.

e (Acidified) permanganate ions, MnO_4^-

f Propan-2-ol

~continue in right column ▲

14 **a** Triglyceride

b

15 **a** $n(C) = n(CO_2) = \dfrac{5.28}{44.01} = 0.119\,97... \text{ mol}$

$m(C) = n \times MM = 0.119\,97... \times 12.01 = 1.4408... \text{ g}$

$n(H) = 2 \times n(H_2O) = 2 \times \dfrac{3.24}{18.016} = 0.359\,68... \text{ mol}$

$m(H) = n \times MM = 0.359\,68... \times 1.008 = 0.362\,557... \text{ g}$

$m(N) = 2.360 - (m(C) + m(H)) = 2.360 - (1.4408... + 0.362\,557...) \text{ g} = 0.5566 \text{ g}$

$n(N) = \dfrac{0.5566}{14.01} = 0.039\,728 \text{ mol}$

C	H	N
0.11997	0.35968	0.039728
3	9	1

The empirical formula is $(C_3H_9N)_n$.

b $PV = nRT$

$n = \dfrac{PV}{RT} = \dfrac{100 \times 1.1888}{8.314 \times (85 + 273.15)} = 0.039\,92... \text{ mol}$

$MM = \dfrac{2.360}{0.03992} = 59.11 \text{ g mol}^{-1}$

c **i** The calculated MM and the mass of the empirical formula are the same, so the molecular formula is C_3H_9N.

Compound X must be an amine as it turned litmus blue. It can be either of the compounds shown below.

Propan-1-amine	Propan-2-amine

12 **1** Label the beakers X and Y.

2 Label two test tubes X and Y.

3 Add 1 mL of each liquid to the appropriate test tube.

4 Test the contents of each test tube with an indicator or litmus paper.

5 The liquid that turns the litmus blue is methanamine and the one that has no effect on litmus is methanamide.

13 **a**

and

b Plasticisers are small molecules inserted between polymer chains so that the chains are further apart. This weakens the intermolecular forces between the chains, making the polymer more flexible.

ii The boiling point of compound X should be tested and compared to literature values of boiling points of propan-1-amine and propan-2-amine. The boiling point of propan-1-amine is slightly higher than the boiling point of propan-2-amine. This is because there are stronger intermolecular forces between the chains when the amine group is in a terminal position; therefore, the chains can line up closer.

d *N*-Propylethanamide

Module eight: Applying chemical ideas

REVIEWING PRIOR KNOWLEDGE PAGE 161

1 Solubility is a measure of the amount of substance that will dissolve in a given volume of water at a specified temperature.

2 Cations are positively charged ions and anions are negatively charged ions.

3 Some combinations of dissolved ions react to produce a solid, called a precipitate.

4 A complete ionic equation shows all ions present in a reaction, while a net ionic equation shows only the reacting species and does not include spectator ions.

5 **a** Reaction proceeds in endothermic direction to decrease temperature, K decreases as less product and more reactant form.

b This is an endothermic reaction so will proceed in the forward direction. K increases.

6 **a** $CaSO_4(s) \rightleftharpoons Ca^{2+}(aq) + SO_4^{2-}(aq)$

b $K_{sp} = [Ca^{2+}][SO_4^{2-}]$

c i Solubility would decrease as reaction would move to left.

 ii No change as the ions added do not react with either of the equilibrium species.

 iii Solubility would decrease as reaction would move to left.

7 a $AgCl(s) \rightleftharpoons Ag^+(aq) + Cl^-(aq)$

$K_{sp} = [Ag^+][Cl^-] = 1.56 \times 10^{-10}$

Let $[Ag^+] = s$ $[Cl^-] = 0.010\,mol\,L^{-1}$

$1.56 \times 10^{-10} = s \times 0.010$

$s = 1.56 \times 10^{-8}$

AgCl will begin to precipitate when $[Ag^+]$
$= 1.56 \times 10^{-8}\,mol\,L^{-1}$

$Ag_2CrO_4(s) \rightleftharpoons 2Ag^+(aq) + CrO_4^{2-}(aq)$

$K_{sp} = [Ag^+]^2[CrO_4^{2-}] = 9.0 \times 10^{-12}$

Let $[Ag^+] = s$ $[CrO_4^{2-}] = 0.0010\,mol\,L^{-1}$

$9.0 \times 10^{-12} = s^2 \times 0.0010$

$s = \sqrt{\dfrac{9.1 \times 10^{-12}}{0.0010}} = 9.5 \times 10^{-5}$

Ag_2CrO_4 will begin to precipitate when $[Ag^+]$
$= 9.5 \times 10^{-5}\,mol\,L^{-1}$

b AgCl should begin precipitating first as there is a lower concentration of Ag^+ needed for precipitation to begin.

c From part a, the $[Ag^+] = 9.5 \times 10^{-5}$ when silver chromate begins to precipitate so:

$[Cl^-] = \dfrac{\left(K_{sp}(AgCl)\right)}{[Ag^+]} = \dfrac{1.56 \times 10^{-5}}{9.5 \times 10^{-5}}$

$= 1.6 \times 10^{-6}\,mol\,L^{-1}$

8 Every measurement is prone to errors or lack of precision, so by repeating a titration until consistent results are achieved, the results should have a greater degree of reliability.

9 a Rinse with the $0.205\,mol\,L^{-1}$ ethanoic acid solution. To rinse with water (or any other solution) would change the concentration of the solution being delivered.

b Rinse the beaker with water. It does not matter if some dilution of the ethanoic acid solution occurs: what is important now is the amount (number of moles) of ethanoic acid present for reaction with added base solution.

c Rinse the burette with the $0.246\,mol\,L^{-1}$ potassium hydroxide solution. To rinse with water (or any other solution) would change the concentration of the solution being delivered.

10 a i –OH functional group

 ii It is an alcohol (a secondary alcohol).

 iii 2-Propanol

b i Double bond

 ii It is an alkene.

 iii 3-Methylbut-1-ene

c i –COOH functional group

 ii It is a carboxylic acid.

 iii 2-Ethylbutanoic acid

d i –COOR

 ii It is an ester.

 iii Methylpropanoate

e i No functional groups

 ii It is an alkane.

 iii 2,3-Dimethylpentane

11 a

1, 2-Dichoropentane

b

3, 4-Dibromohexane

c

Sodium methanoate

d

Propyl ethanoate

9780170449656

12 a, b Choosing alcohols with four carbon atoms in their molecules (i.e. isomers):

i
Butan-1-ol

ii
Butan-2-ol

iii
2-Methylpropan-2-ol

c Acidified dichromate ion solution:

  i will oxidise the primary alcohol to an aldehyde, and then to a carboxylic acid (in this case, butanal, and then butanoic acid)

  ii will oxidise the secondary alcohol to a ketone (in this case, 2-butanone)

  iii will not oxidise tertiary alcohols.

13 a $H_2SO_4(aq) + CaCO_3(s) \rightarrow CaSO_4(s) + H_2O(l) + CO_2(g)$

$m(CaCO_3) = 3.40\,g$ $MM = 100.09\,g$

$n(CaCO_3) = \dfrac{3.40}{100.09} = 0.03397...\,mol$

$n(H_2SO_4) = c \times V = 1.00 \times 0.1 = 0.100\,mol$

Reaction ratio $H_2SO_4 : CaCO_3 = 1 : 1$

Limiting reagent is $CaCO_3$.

b Excess $H_2SO_4 = 0.100 - 0.033\,97 = 0.0660\,mol$

c $n(CO_2) = 0.033\,97\,mol$ $MM = 44.01\,g$

$m(CO_2) = 0.033\,97 \times 44.01 = 1.195\,g = 1.20\,g$

Chapter 14: Analysis of inorganic substances

WS 14.1 PAGE 165

1 a

	Cl^-	SO_4^{2-}	CO_3^{2-}	OH^-
Ca^{2+}	✓	✓	✓	✓
Mg^{2+}	✓	✗	✓	✓
Cu^{2+}	✗	✗	✓	✓
Ba^{2+}	✓	✓	✓	✗
Ag^+	✓	✓	✓	✓

Cl^- does not form precipitates with Ca^{2+}, Mg^{2+}, Ba^{2+}, so rows 1, 2 and 4 in the Cl^- column are incorrect.

b There are two possibilities. Either the Cl^- solution is incorrectly labelled or it is contaminated. In either instance, it is likely that the solution/contaminate is CO_3^{2-} as this is the only ion involved in the testing that precipitates with all the cations.

2 a The investigation should give an aim that relates to identifying if the solution has been mislabelled, i.e. it only contains or if it has been contaminated, i.e. contains both Cl^- and CO_3^{2-} ions.

Outlined below is a process that could be used as the basis of the investigation. The equipment and method should be consistent with the outline provided below.

In order to determine if there is CO_3^{2-} present, add a few drops of acid to the solution. If bubbling occurs, it indicates the presence of carbonate ions.

To determine if contamination occurred, the CO_3^{2-} ions need to be removed and the remaining solution tested for Cl^-. The CO_3^{2-} can be removed from a sample of the solution by:

▶ reacting the CO_3^{2-} ions in the solution with an excess of acid, HNO_3 (cannot use H_2SO_4 or HCl as these would introduce ions which could produce precipitation), to produce carbon dioxide and water.

or

▶ adding an excess of any of Ca^{2+}, Mg^{2+}, Cu^{2+} or Zn^{2+} cations as a nitrate solution, which will precipitate out the CO_3^{2-} ions in the solution. The solution could then be filtered to remove the precipitate.

Once the CO_3^{2-} ions have been removed, the solution would then be tested to see if there are Cl^- ions. This could be done by adding $AgNO_3$ to some of the solution.

b If a precipitate forms, the solution contains Cl^- ions and the original solution had been contaminated. If no precipitate forms, then it is the CO_3^{2-} solution which has been incorrectly labelled as Cl^-.

3 The equations given should match the investigation. For example:

$CO_3^{2-}(aq) + 2H^+(aq) \rightarrow CO_2(g) + H_2O(l)$

$Ca^{2+}(aq) + CO_3^{2-}(aq) \rightarrow CaCO_3(s)$

$Ag^+(aq) + Cl^-(aq) \rightarrow AgCl(s)$

WS 14.2 PAGE 167

1 a Fe^{2+}, Fe^{3+}, Cu^{2+}, Pb^{2+}, Ca^{2+}

b Precipitation reactions and flame tests

c All the cation samples must be soluble in water.

d Distilled water should be used as this ensures there are no other ions present which could contaminate the samples.

e **i** Pb^{2+}

  ii When excess OH^- is added, the lead(II) hydroxide binds to two more OH^- ions to form a complex ion that is soluble.

2 a $A = OH^-$; $B = CH_3COO^-$; $C = Cl^-$; $D = SO_4^{2-}$

b **i** $2Ag^+(aq) + SO_4^{2-}(aq) \rightarrow Ag_2SO_4(s)$

  ii $Pb^{2+}(aq) + 2OH^-(aq) \rightarrow Pb(OH)_2(s)$

c It is possible to separate the anions; however, the process would involve forming precipitates so compounds of the anions would be the result of the separation. It would not be possible to obtain the actual anion in its ionic form.

  1 Add a solution containing Mg^{2+} ions (e.g. $Mg(NO_3)_2$) to the mixture of anions and filter to remove the precipitate, which is MgA ($Mg(OH)_2$).

  2 Add a solution containing Ba^{2+} ions (e.g. $Ba(NO_3)_2$) to the filtrate, then filter to remove the precipitate, which is BaD ($BaSO_4$).

  3 Add a solution containing Pb^{2+} ions (e.g. $Pb(NO_3)_2$) to the filtrate, then filter to remove the precipitate, which is PbC ($PbCl_2$).

  4 The remaining filtrate will contain CH_3COO^-. This could be removed by adding a solution containing Ag^+ (e.g. $AgNO_3$) and filtering to obtain the precipitate $AgCH_3COO$.

3 a HPO_4^{2-} is a very weak acid as can be seen by the small K_a value; therefore, there will be only a very small concentration of PO_4^{3-} ions present in solution. Adding ammonia, which does not precipitate with any of the cations used for testing, removes H_3O^+ ions, which shifts the equilibrium to the right resulting in a higher concentration of PO_4^{3-} ions.

b Pb^{2+} ions would be the best to use as the K_{sp} for lead phosphate is the smallest of all the phosphate ion compounds that are given. Therefore, the precipitate with phosphate ions is the least soluble, so lower concentrations could be detected.

c $Pb_3(PO_4)_2(s) \rightleftharpoons 3Pb^{2+}(aq) + 2PO_4^{3-}(aq)$ $K_{sp} = 8.0 \times 10^{-43}$

Let $[Pb^{2+}] = 3s$; therefore, $[PO_4^{3-}] = 2s$

$K_{sp} = [Pb^{2+}]^3 \times [PO_2^{3-}] = 8.0 \times 10^{-43}$

$(3s)^3 \times (2s)^2 = 8.0 \times 10^{-43}$

$108s^5 = 8.0 \times 10^{-43}$

$s = \sqrt[5]{\dfrac{8.0 \times 10^{-43}}{108}} = 1.49... \times 10^{-9}$

$[PO_4^{2-}] = 2s = 2.98... \times 10^{-9} = 3.0 \times 10^{-9}\,mol\,L^{-1}$

d $HPO_4^{2-}(aq) + H_2O(l) \rightleftharpoons PO_4^{3-}(aq) + H_3O^+(aq)$

Assume $[PO_4^{3-}] = [H_3O^+]$ as dissociation is $1:1$

$[H_3O^+] = 3.0 \times 10^{-9}\,mol\,L^{-1}$

$pH = -\log(3.0 \times 10^{-9}) = 8.532 = 8.5$

e Other precipitates of phosphate ions are more soluble; therefore, a higher concentration of phosphate ions would be needed for the precipitate to form. This means the equilibrium would need to shift more to the right so the pH would need to be lower since the $[H_3O^+]$ must be greater.

WS 14.3 PAGE 170

1 a Using the same brand should help control variables as it would be expected that similar ingredients and production processes would be used for both types. This should assist in improving the validity of the investigation.

b i As the student has information about the sodium content and that soy sauce contains a high amount of salt, they would be testing and measuring the chloride ion. They would have most likely assumed that as Na^+ and Cl^- exist in a $1:1$ ratio in sodium chloride, there would be equal amounts of sodium ions and chloride ions in the samples.

ii Even though the soy sauce sample is diluted, it will still have a light brown colour. In Mohr's method, the end point colour change produces a red-brown precipitate of silver chromate and this colour change may be difficult to identify in a brown solution. The silver chloride precipitate may also have a brownish tinge due to other substances in the soy sauce. Volhard's method has two advantages for this analysis. One is that the precipitate is filtered off before the titration is done and the second is that the colour change at the end point is a distinctive blood red colour due to the formation of $Fe(SCN)^{2+}$.

c Using distilled water would ensure no other ions were introduced into the sample. Using a pipette and a volumetric flask to dilute the sample ensures measurements are as accurate as possible.

d Regular soy sauce   Average = 26.0 mL
Low-salt soy sauce   Average = 33.9 mL

e i $Ag^+(aq) + SCN^-(aq) \rightarrow AgSCN(s)$

For regular soy: $n(Ag^+) = n(SCN^-) = cV = 0.020 \times 0.0260 = 5.20 \times 10^{-4}\,mol$ in 20 mL

For low-salt soy: $n(Ag^+) = n(SCN^-) = cV = 0.020 \times 0.0339 = 6.78 \times 10^{-4}\,mol$ in 20 mL

ii Total volume of solution for each type of soy = 210 mL

Regular soy $n(Ag^+) = 5.20 \times 10^{-4} \times \dfrac{210}{20} = 5.46 \times 10^{-3}\,mol$

Low-salt soy $n(Ag^+) = 6.78 \times 10^{-4} \times \dfrac{210}{20} = 7.119 \times 10^{-3}\,mol$

iii $Ag^+(aq) + Cl^-(aq) \rightarrow AgCl(s)$

$n(Ag^+)$ added to 200 mL of each type of soy $= 1.00 \times 0.010 = 0.010\,mol$

Regular soy $n(Ag^+)$ reacted $= 0.010 - 5.46 \times 10^{-3}$
$= 4.54 \times 10^{-3}\,mol = n(Cl^-)$

Low-salt soy $n(Ag^+)$ reacted $= 0.010 - 7.119 \times 10^{-3}$
$= 2.88 \times 10^{-3}\,mol = n(Cl^-)$

iv $n(Cl^-)$ in 500 mL = $n(Cl^-)$ in original 5 mL (solution was diluted but n constant) = $n(Na^+)$

Regular soy: $n(Cl^-) = 4.54 \times 10^{-3} \times \dfrac{500}{200} = 0.011\,35\,mol = n(Na^+)$

$m(Na^+) = 0.011\,35 \times 22.99 = 0.2609\,g$ in 5 mL

$= \dfrac{0.2609}{5} = 0.052\,18\,g\,mL^{-1}$

Low-salt soy: $n(Cl^-) = 2.88 \times 10^{-3} \times \dfrac{500}{200} = 0.007\,20\,mol$
$= n(Na^+)$

$m(Na^+) = 0.007\,20 \times 22.99 = 0.166\,g$ in 5 mL

$= \dfrac{0.166}{5} = 0.0331\,g\,mL^{-1}$

f

Sample	Manufacturer value mg/100 mL	Experimental value mg/100 mL
Regular soy sauce	6833	5218
Low-salt soy sauce	3560	3310

The experimental values are both lower than the values provided by the manufacturer. There may have been some inaccurate measurement of the samples and in the addition of $AgNO_3$. Due to the colour of the soy solutions, it may have been difficult to accurately determine the end point. It was also assumed that all the sodium came from sodium chloride; however, there may have been sodium from another compound that was not measured, resulting in a lower amount of sodium being calculated.

g Observation of the precipitate showed it to be a brownish colour rather than the typical white associated with $AgCl$. This means that other compounds in the soy may have become trapped in the precipitate. In using gravimetric analysis, these compounds would have been included in the mass, resulting in a larger mass and hence a larger amount of sodium being calculated.

2 a $Ba^{2+}(aq) + SO_4^{2-}(aq) \rightarrow BaSO_4(s)$

$n(BaSO_4) = \dfrac{5.82}{(137.3 + 32.07 + 4 \times 16.00)} = 0.0249...\,mol$

$n(SO_4^{2-}) = n(BaSO_4) = 0.0249\,mol$

b $Al_2(SO_4)_3(aq) \rightarrow 2Al^{3+}(aq) + 3SO_4^{2-}(aq)$

$n(Al_2(SO_4)_3) = \dfrac{1}{3}n(SO_4^{2-}) = \dfrac{0.0249...}{3} = 0.008\,312...\,mol$

$m(Al_2(SO_4)_3) = 0.008\,312... \times (342.3) = 2.845...\,g = 2.85\,g$

c $m(\text{impurity}) = 3.00 - 2.85 = 0.15\,g$

$\%\,(\text{impurity}) = \dfrac{0.15}{3.00} \times 100 = 5\%$

WS 14.4 **PAGE 174**

1 a

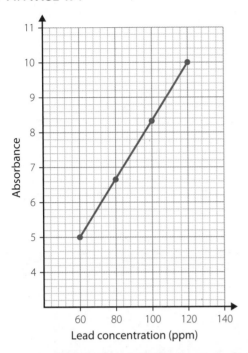

b

	Sample	Absorbance	Lead concentration (ppm)
1	Yellow	10.25	123
2	White	4.50	55
3	Red	8.90	107
4	Blue	7.85	95

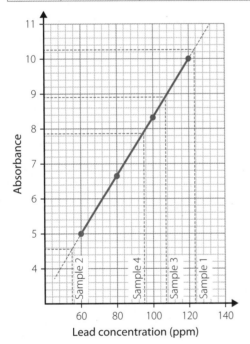

c $90\,mg\,L^{-1} = 90\,ppm$ so only sample 2, the white paint, is allowed.

d Only the values for samples 3 and 4 lie within the data used to produce the calibration curve, so the results for these two samples should be considered valid. The results for samples 1 and 2 lie outside the calibration curve data and have been obtained by extrapolating the calibration curve, assuming the trend continues. There is no evidence to support this, so the validity of the concentration of these samples is questionable.

2 a Absorbance $0.50 = 18\,mg\,L^{-1}$; absorbance $0.60 = 24.9\,mg\,L^{-1}$
Min amount lost $= 18 \times 1312 \times 365 = 8\,619\,840\,mg = 8.6\,kg$
Max amount lost $= 24.9 \times 1312 \times 365 = 11\,924\,112\,mg = 12\,kg$

b Absorbance $0.20 = 5.2\,mg\,L^{-1}$
Amount lost $= 5.2 \times 1312 \times 365 = 2490176\,mg = 2.5\,kg$
The reduction of loss in the smelter would be between 6.1 and 9.5 kg.

3 a

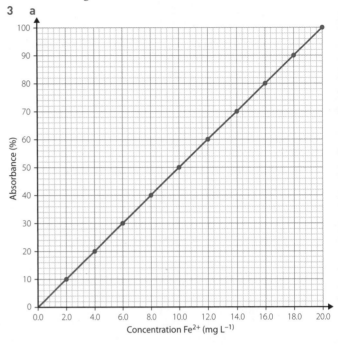

b i $19\,mg\,L^{-1}$

ii Blood concentration $= 19 \times 10 = 190\,mg\,L^{-1}$

iii Total iron $= 190 \times 5.5 = 1045\,mg = 1.045\,g$

c Yes, the patient should be concerned. They have less than half the amount of iron that is needed in a healthy adult.

Chapter 15: Analysis of organic substances

WS 15.1 **PAGE 178**

1 a A = alcohol, B = alkane, C = alkene, D = carboxylic acid, E = alcohol

b Taking an IR or NMR spectrum of each of the compounds would definitely confirm whether they were the same compound. If the spectra were identical, then A and E would be the same compound.

2 a Alcohol and alkene

b i The bromine water would turn from brown to colourless.

ii

$$H-\overset{\overset{\displaystyle H}{|}}{\underset{\underset{\displaystyle H}{|}}{C}}-\overset{\overset{\displaystyle H}{|}}{\underset{\underset{\displaystyle H}{|}}{C}}-\overset{\overset{\displaystyle H}{|}}{C}=C-\overset{\overset{\displaystyle H}{|}}{\underset{\underset{\displaystyle H}{|}}{C}}-OH + Br_2 \rightarrow H-\overset{\overset{\displaystyle H}{|}}{\underset{\underset{\displaystyle H}{|}}{C}}-\overset{\overset{\displaystyle H}{|}}{\underset{\underset{\displaystyle H}{|}}{C}}-\overset{\overset{\displaystyle H}{|}}{\underset{\underset{\displaystyle Br}{|}}{C}}-\overset{\overset{\displaystyle Br}{|}}{\underset{\underset{\displaystyle H}{|}}{C}}-\overset{\overset{\displaystyle H}{|}}{\underset{\underset{\displaystyle H}{|}}{C}}-OH$$

2, 3-dibromopentan-1-ol

c i Student M's suggestion would not confirm the presence of the alcohol group as alkenes also decolourise permanganate. Student N's suggestion would confirm the presence of the alcohol as the proposed reaction would produce an ester that could be identified by a new odour and insoluble layer in the cold water.

ii As compound A is a primary alcohol, addition of acidified permanganate as suggested by student M would result in the −OH functional group being oxidised to an aldehyde, which would then be oxidised to a carboxylic acid. Student N's suggestion would result in the production of an ester.

1 A mass spectrometer is designed to accelerate and deflect positively charged particles. The positively charged particles are accelerated by an electric field and then deflected by a magnetic field. The amount of deflection depends on the strength of the magnetic field as well as the mass and charge of the particles. As the free radical is neutral, it is not affected by either the electric or magnetic field so cannot be detected.

2 a i Peak of the parent molecular ion is at 58.

   ii C = 12, H = 1; general formula for alkanes is C_nH_{2n+2}

   Try C_4H_{10} $4 \times 12 + 1 \times 10 = 58$

   Parent molecular ion is C_4H_{10}.

   b i, ii

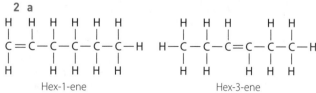

Butane   Methylpropane

   c Addition of H_2 occurs across the double bond. Butane would be produced from butene and methylpropane would be produced from methylpropene. Butene has two possible isomers – but-1-ene and but-2-ene. However, methylpropene has only one isomer; therefore, the unknown alkane must be butane.

   d Base peak is 43 = $C_3H_7^+$

   42 is the base peak less 1 H = $C_3H_6^+$

   41 is base peak less 2 H = $C_3H_5^+$

   29 = $C_2H_5^+$

   28 is peak at 29 less 1 H = $C_2H_4^+$

   27 is peak at 29 less 2 H = $C_2H_3^+$

   15 = CH_3^+

3 a The sample is contaminated. When comparing the spectrum there are lines at 88, 70, 61 and 45 in the sample spectrum which do not exist in the reference spectrum.

   b Comparing the data in the table with the mass spectrum of the contaminated sample, the presence of lines at 70, 61 and 45 confirm the contaminant as ethyl ethanoate. While there are 4 lines at 88, 59, 29 and 15 that correspond to methyl propanoate, the absence of lines at 57 and 27 indicate this compound is not present.

1 ¹H NMR spectroscopy uses ¹H (or protons) to determine the number of different environments for the hydrogens, the number of hydrogens in each environment, the type of proton, and, if it is a high-resolution spectrum, the number of hydrogens on the adjacent carbon atoms. The ¹³C NMR spectroscopy uses the less abundant ¹³C isotope to determine the number of different carbon environments in the molecule and the type of carbon present. Therefore, a ¹H NMR spectrum can provide a lot more information than a ¹³C NMR spectrum.

2 a

H H H H H H   H H H   H H
C=C—C—C—C—C—H   H—C—C—C=C—C—C—H
H   H H H H   H H   H H H

Hex-1-ene   Hex-3-ene

   b Spectrum A is hex-1-ene and spectrum B is hex-3-ene. As can be seen in the diagrams of the structures, hex-1-ene has six different carbon environments, while hex-3-ene has only three different carbon environments due to its symmetry.

c i There would be three peaks because there are three different hydrogen environments:

   ▸ the CH_3 group adjacent to a CH_2 group

   ▸ the CH_2 group adjacent to both the CH_3 group and the H on the double-bonded C.

   ▸ the H on the double-bonded C adjacent to another H on the other double-bonded C and the CH_2 group.

   ii The peak for the CH_3 group will split into a triplet due to the presence of the neighbouring CH_2 group. The peak for the CH_2 group will split into a quintet due to the neighbouring CH_3 group and single H (3H + 1H + 1). It would be expected that the peak for the individual H on the double-bonded C would split into a quartet (adjacent to CH_2 and CH). However, as the H on the adjacent CH is in exactly the same chemical environment, it will not contribute to a split so the peak splits into a triplet not a quartet.

3 a i

C	H	O
52.3%	13.1%	34.6%
$\dfrac{52.3}{12.01} = 4.32$	$\dfrac{13.1}{1.008} = 13.0$	$\dfrac{34.6}{16.0} = 2.16$
$\dfrac{4.32}{2.16} = 2$	$\dfrac{13.0}{2.16} = 6.02$	$\dfrac{2.16}{2.16} = 1$

   Empirical formula = C_2H_6O

   ii The weight of the empirical unit is 46.1 so the molecular formula is the same as the empirical formula C_2H_6O.

   b The molecular formula of X is C_2H_6O. The structure of isomers with this molecular formula are shown below.

H H   H   H
H—C—C—OH   H—C—O—C—H
H H   H   H

   In the ¹H NMR spectrum of X, there are three peaks, indicating three different hydrogen environments. The first structure would exhibit three different H environments, while the second structure would only exhibit one as the environment for all the H is the same.

   The peak at 1.2 is a triplet, which indicates it is adjacent to a CH_2 group, and the position and intensity would indicate it is a CH_3 group. The quartet peak at 3.7 indicates an adjacent CH_3 group and the shift is consistent with a CH_2 group on the same C as an OH group. The single peak at 2.6 would be the OH group as the there is no interaction between the H on an OH and any neighbouring H.

   Based on the analysis, compound X is ethanol.

   Ethanol undergoes a dehydration reaction when heated with sulfuric acid, so this reaction would produce ethene.

   $CH_3CH_2OH(aq) \rightarrow CH_2CH_2(l) + H_2O(l)$

   Therefore, compound Y should be ethene. The ¹³C NMR spectrum for compound Y shows only one peak at 123 ppm. The typical shift for –C=C– is between 90 and 150. Ethene is a symmetrical molecule and there is only 1 carbon environment, so the data is consistent with the conclusion that compound Y is ethene.

   Ethene undergoes an addition reaction with HCl to form chloroethane.

   $CH_2CH_2(l) + HCl(aq) \rightarrow CH_3CH_2Cl(aq)$

   Therefore, compound Z should be chloroethane. The ¹H NMR spectrum for compound Z shows two peaks, indicating two H environments. The quartet peak at 3.5 indicates the presence of an adjacent CH_3 group and the 3.5 shift value is consistent with the shift of a H, which is part of CH_2-Cl combination.

The triplet at 1.5 indicates an adjacent CH_2 group. The 2:3 ratio of the peaks is also consistent with the presence of a CH_3 and a CH_2 group. Therefore, analysis of the data is consistent with the conclusion that compound Z is chloroethane.

WS 15.4 PAGE 186

1 Both techniques use wavelengths of the electromagnetic spectrum. UV–vis uses the UV and visible part of the spectrum, while IR uses the IR part of the spectrum. UV–visible relies on the electrons in the molecule absorbing energy and moving to a higher energy state. The amount of energy absorbed is recorded. IR relies on energy increasing the vibrations of the molecules. It detects functional groups present and can confirm when they are not present. The amount of energy transmitted is recorded.

Both techniques are non-destructive where the light is shone on the samples. The identity of substances can be determined by comparing the spectra of a pure sample to those of the compound. Both are relatively cheap and easy techniques.

2 a Propan-1-ol will be oxidised to propanal, then to propanoic acid.

Propan-2-ol will be oxidised to propanone.

b Product B spectrum is that of propanoic acid. The spectrum shows a broad band at $3000\ cm^{-1}$, which corresponds to the O–H bond of an acid, the peak at around $1700\ cm^{-1}$ corresponds to a C=O bond, and the peak at $1200\ cm^{-1}$ corresponds to the C–O bond all of which are present in a carboxylic acid.

Product A spectrm is that of propanone. The spectrum shows a peak at $1700\ cm^{-1}$, which corresponds to a C= O bond. However, there is no broad band at 2500–$3000\ cm^{-1}$, which indicates an absence of an O–H bond.

c Both reactions went to completion. If any reactant alcohol was present in the final mixture, the spectra would have shown a broad peak between 3230 and $3550\ cm^{-1}$. As neither of the spectra show this peak, it can be concluded that there was no alcohol in the final mixture and therefore both reactions went to completion.

3 a i λ_{max} is the wavelength at which the greatest amount of absorption occurs.

ii $\lambda_{max} = 470$ nm

b Using UV–visible spectroscopy allows the concentration of a compound which absorbs light at a particular wavelength to be determined. Using 470 nm means any lycopene present would absorb the light. As the number of lycopene molecules (concentration) increases, the absorption would also increase.

4 a

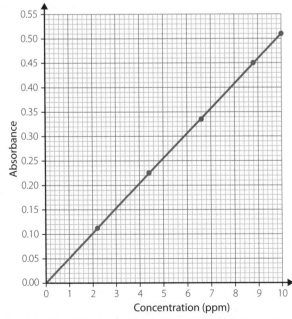

b The concentration of carmine in the diluted sample is 5.3 ppm. The sample had been diluted by a factor of 10 so the concentration in the original sample is 53 ppm. The original sample of 100 mL contains 53 ppm.

c The daily limit is $5\ mg\,kg^{-1}$, which is 5 ppm. The person would have consumed 5×53 ppm = 265 ppm. This means they would have exceeded the daily limit by 260 ppm.

WS 15.5 PAGE 189

1 a

C	H	O
70.59%	13.72%	15.69%
$\dfrac{70.59}{12.01} = 5.878$	$\dfrac{13.72}{1.008} = 13.61$	$\dfrac{15.69}{16.0} = 0.9806$
$\dfrac{5.878}{0.0986} = 5.99$	$\dfrac{13.61}{0.9806} = 13.88$	$\dfrac{0.9806}{0.9806} = 1$
6	14	1

Empirical formula is $C_6H_{14}O$.

The mass spectrum information shows the relative molecular mass is 102 as this is the molecular peak.

MW of $C_6H_{14}O = 6 \times 12.01 + 14 \times 1.008 + 16 = 102.172$; therefore, the empirical formula is also the molecular formula.

The IR spectrum shows a broad peak at $3350\ cm^{-1}$, which indicates the presence of –OH; therefore, the compound is an alcohol.

As it is an alcohol and unbranched it could be:

Hexan-1-ol $CH_3CH_2CH_2CH_2CH_2CH_2OH$

Hexan-2-ol $CH_3CH_2CH_2CH_2CH_2 (OH)CH_3$

Hexan-3-ol $CH_3CH_2CH_2CH_2(OH)CH_2CH_3$

b i Hexan-1-ol $CH_3CH_2CH_2CH_2CH_2CH_2OH$
5 peaks in ^1H NMR   6 peaks in ^{13}C NMR

Hexan-2-ol $CH_3CH_2CH_2CH_2CH_2(OH) CH_3$
5 peaks in ^1H NMR   6 peaks in ^{13}C NMR

Hexan-3-ol $CH_3CH_2CH_2CH_2(OH)CH_2CH_3$
4 peaks in ^1H NMR   6 peaks in ^{13}C NMR

ii ^{13}C NMR would not be useful as all three of the isomers show six peaks. ^1H NMR would be useful to identify hexan-3-ol but it would still not be useful to distinguish between hexan-1-ol and hexan-2-ol as both these molecules have five peaks.

c Chemical testing could be used to distinguish between the primary and secondary alcohols. Chemical tests that could be used are:

 ‣ add a rice-sized grain of Na metal to each of the samples and compare the rates of bubbling; primary alcohols react faster than secondary alcohols

 ‣ add acidified potassium permanganate solution to each of the samples; this would result in producing a carboxylic acid in the case of a primary alcohol and a ketone in the case of a secondary alcohol, then add solid Na_2CO_3 and bubbling identifies the presence of an acid thus the primary alcohol.

This would still not distinguish between the two secondary alcohols.

2 Mass spectra:

1,1-Dichloropropene: CCl_2=$CHCH_3$ molecular mass of 111, spectrum B

Propanone: $CH_3C(O)CH_3$ molecular mass of 58, spectrum A

Propan-1-ol: $CH_3CH_2 CH_2OH$ molecular mass of 60, spectrum C

¹³C NMR spectra:

1,1-Dichloropropene: $CCl_2=CHCH_3$ has three different C environments, so has three peaks, C=C shift 90–150 ppm, C-Cl 10–70 ppm, spectrum D

Propanone: $CH_3C(O)CH_3$ has two different C environments so has two peaks, C=O shift at 190–220 ppm, C(O)–C shift 20–50 ppm, spectrum E

Propan-1-ol: $CH_3CH_2CH_2OH$ has three different C environments, so has three peaks, CH_3 shift 8–25 ppm, C–O with a shift 50–90 ppm, spectrum F

IR spectra:

1,1-Dichloropropene: $CCl_2=CHCH_3$; No spectrum provided; however, would expect C=C band at 1620–1680 cm^{-1}

Propanone: $CH_3C(O)CH_3$; C=O band at 1700 cm^{-1}, no OH band spectrum H

Propan-1-ol: $CH_3CH_2 CH_2OH$; Broad OH band 3300 cm^{-1}, C–O band at 1000 cm^{-1}, spectrum G

Chapter 16: Chemical synthesis and design

WS 16.1 PAGE 194

1 a

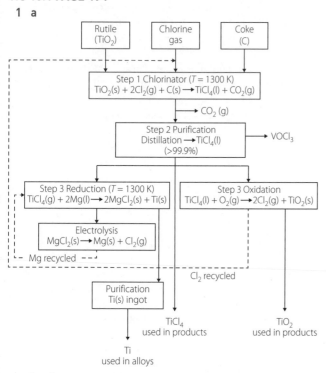

b i Rutile, chlorine gas, coke

ii Ti metal, $TiCl_4$, TiO_2

iii Mg, Cl_2

iv CO_2, $VOCl_3$

c i Chlorine is a halogen and is highly reactive. When chlorine reacts, it forms the Cl$^-$ ion, which has a completed outer shell, making it more chemically stable. Reducing the Cl$^-$ ion requires removal of an electron, resulting in a higher energy, less stable configuration. This process would occur slowly.

ii The reduction of the Cl$^-$ ion requires a significant energy input. This implies the reaction has a high activation energy. Increasing the temperature will increase the energy of the reactants, resulting in more reactants with sufficient energy to overcome the activation energy barrier. It will also increase the rate at which collisions occur so both these factors will lead to an increase in the reaction rate.

iii The reaction is between $TiCl_4(g)$ and Mg(l). The reaction is occurring between heterogeneous states, which implies it is occurring at the surface of Mg(l) and the solids produced are being deposited. Increasing the temperature too much higher would lead to the Mg(l) → Mg(g) and this may have the effect of reducing reaction rate due to fewer collisions between gaseous molecules.

2 a i The temperature is above the boiling point of the two reactants and the products so all would be in a gaseous state.

ii As all the species are in a gaseous state, increasing pressure would result in an equivalent increase in concentration of all species. However, according to Le Chatelier's principle, the equilibrium would shift to the right in order to partially counteract the pressure increase as the right-hand side has less gaseous molecules. The yield would be increased due to the production of more product.

b i As the reaction is exothermic, decreasing the temperature would shift the equilibrium to the right (exothermic direction) as the reaction acts to counter the imposed change. This would result in more product being made.

ii Although lowering the temperature would increase the yield, the rate of the reaction would also be reduced as the molecules would have lower kinetic energy and fewer molecules would be able to overcome the activation energy barrier. Therefore, the temperature used is a trade-off between high yield and lower rate.

c i Isopropylbenzene is an intermediate. It is formed in step 1 and used as a reactant in step 2.

ii Chemists choose a reaction pathway based on many factors, including the conditions required for the reaction. It is likely that the two steps were chosen in order to minimise the energy and pressure requirements, maximise yields and carry out the reactions as cheaply and safely as possible.

iii

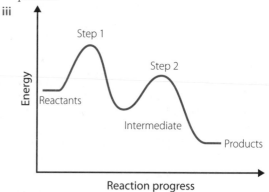

WS 16.2 PAGE 198

1 a The efficiency of a process, percentage yield, can be determined by comparing actual yield with theoretical yield. Theoretical yield is the amount of product that would be obtained in a reaction if all the reactants were used up and the reaction went to completion. It is based on the assumption that the reaction is 100% efficient. Yield is the actual amount of product that is obtained from a reaction.

b The percentage yield is a measure of the quantity of a product while purity is a measure of the quality of product. The level of purity necessary in a product is determined by its use.

9780170449656

2 a $HCl(aq) + NaOH(aq) \rightarrow NaCl(aq) + H_2O(l)$

$HCl(aq) + NH_3(g) \rightarrow NH_4^+(aq) + Cl^-(aq)$

$(NH_4)_2SO_4(aq) + 2NaOH(aq) \rightarrow Na_2SO_4(aq) + 2NH_3(g) + 2H_2O(l)$

Reaction between NaOH solution and excess HCl:

$n(NaOH) = cV = 0.100 \times 0.0455 = 0.004\,55\,mol$

$n(HCl) = n(NaOH) = 0.004\,55\,mol$

Initial $n(HCl) = 0.100 \times 0.050 = 0.0050\,mol$

Reaction between HCl and NH_3:

$n(HCl)$ reacted with $NH_3 = 0.0050 - 0.004\,55 = 0.000\,45\,mol$

$n(NH_3) = n(HCl) = 0.000\,45\,mol$

However, this was in a 25 mL sample.

Total sample volume was 250 mL, so $n(NH_3) = 10 \times 0.000\,45 = 0.0045\,mol$

$n((NH_4)_2SO_4) = ½\,n(NH_3) = \dfrac{0.0045}{2} = 0.002\,25\,mol$

Mass $(NH_4)_2SO_4 = 132 \times 0.002\,25 = 0.297\,g$

$\%(NH_4)_2SO_4 = \dfrac{0.297}{3.00} \times 100 = 9.9\%$

b Mass of $(NH_4)_2SO_4$ in 3.00 g $= 3.00 \times 0.24 = 0.72\,g$

c % yield $= \dfrac{0.297}{0.72} \times 100 = 41.3\%$

3 $n(propene) = \dfrac{15}{42} = 0.357\,mol$

$n(glycerol) = \dfrac{1}{92} = 0.0109\,mol$

% yield $= \dfrac{0.0109}{0.357} \times 100 = 3.05 = 3.1\%$

4 Reactions

$CaCO_3(s) + 2HCl(aq) \rightarrow CaCl_2(aq) + H_2O(l) + CO_2(g)$

$HCl(aq) + NaOH(aq) \rightarrow NaCl(aq) + H_2O(l)$

$n(NaOH) = 0.0500 \times 0.0232 = 0.001\,16\,mol$

$n(HCl) = 0.001\,16\,mol$ (in 25 mL)

In 250 mL:

$n(HCl) = 10 \times 0.00116 = 0.0116\,mol$

Original amount of HCl added to $CaCO_3 = 1.50 \times 0.025 = 0.0375\,mol$

Amount of HCl reacted with $CaCO_3$ = amount of HCl added to $CaCO_3$ sample – amount remaining after reaction with NaOH = $0.0375 - 0.0116 = 0.0256\,mol$

From balanced equation $CaCO_3 : HCl = 1 : 2$

$n(CaCO_3) = \dfrac{0.0256}{2} = 0.0128\,mol$

$m(CaCO_3) = 0.0128 \times 100.1 = 1.28\,g$

% purity $= \dfrac{1.28}{1.32} \times 100 = 97\%$

5 a Spectrum A is propan-2-ol. It has a single –OH peak at 2.2, a doublet at 1, which corresponds the CH_3 attached to the –CH, and a septet peak, which corresponds to the central CH, which is adjacent to two CH_3 groups.

Spectrum B is ethanol. It has a single –OH peak at 2.6, a quartet peak that corresponds to the CH_2 group adjacent to a CH_3 and a triplet which corresponds to a CH_3 adjacent to a CH_2.

b i $n(S_2O_3^{-2}) = 1 \times 0.0316 = 0.0316\,mol$

ii $n(I_2) = ½\,n(S_2O_3^{-2}) = \dfrac{0.0316}{2} = 0.0158\,mol$

iii $n(Cr_2O_7^{2-}) = \dfrac{0.0158}{3} = 0.005\,266...\,mol$

iv $n(Cr_2O_7^{2-})$ reacted $= n(Cr_2O_7^{2-})$ added initially $- n(Cr_2O_7^{2-})$ in excess

$n(Cr_2O_7^{2-})$ added initially $= 0.4 \times 0.020 = 0.0080\,mol$

$n(Cr_2O_7^{2-})$ reacted $= 0.0080 - 0.005\,27 = 0.002\,73\,mol$

v $n(alcohol) = n(Cr_2O_7^{2-}) \times 3 = 0.002\,73 \times 3 = 0.0082\,mol$

vi $n(alcohol)$ in 20 mL $= 0.0082\,mol$

$n(alcohol)$ in 250 mL $= 0.0082 \times \dfrac{250}{20} = 0.1025\,mol$

$m(alcohol) = 0.1025 \times 46.068 = 4.72\,g$

vii $V(alcohol) = \dfrac{4.72}{0.785} = 6.015\,mL$

% alcohol $= 60.2\%$

c Both claims were incorrect. The manufacturer claimed the product was 70% alcohol; however, analysis showed it to be only 60%, and therefore was not as effective against viruses. The media report was also incorrect as the reported value of 30% was half the actual amount.

MODULE EIGHT: CHECKING UNDERSTANDING PAGE 202

1 C **2** C **3** D **4** A **5** B

6 D **7** A **8** C **9** B **10** C

11 There are many possible pathways that could be used. An example is provided.

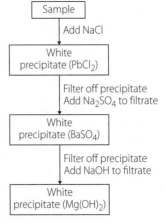

12 $n(S_2O_3^{2-}) = 0.200 \times 0.0262 = 0.005\,24\,mol$

$n(I_2) = \dfrac{0.005\,24}{2} = 0.002\,62\,mol$

$m(I_2) = 0.002\,62 \times (126.9 \times 2) = 0.6649...\,g$

% purity $= \dfrac{0.6649...}{0.80} \times 100 = 83.1\%$

13 a Propanone is a ketone so the peak around $1700\,cm^{-1}$ indicates the presence of a $—C=O$ group.

b The presence of the broad band above $3000\,cm^{-1}$ indicates the presence of another organic compound in the layer. This is most likely an alcohol as a band at $3300\,cm^{-1}$ is typical of an –OH group.

14 a In colourimetry, a filter of a complementary colour to that being measured is used. As the solution appears yellow, only yellow light passes through the solution so blue is the complementary colour that is most strongly absorbed.

b

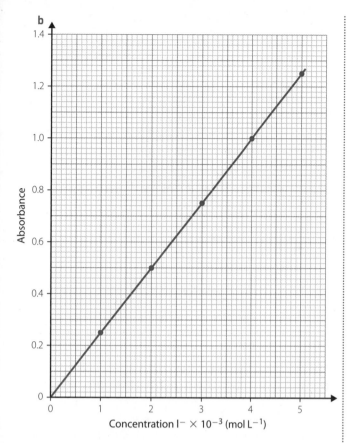

c i $PbI_2(s) \rightleftharpoons Pb^{2+}(aq) + 2I^-(aq)$

$K_{sp} = [Pb^{2+}][I^-]^2$

ii An absorbance of 0.68 is produced by $[I^-] = 2.7 \times 10^{-3} \, mol \, L^{-1}$

$[Pb^{2+}] = \dfrac{[I^-]}{2} = \dfrac{2.7 \times 10^{-3}}{2} = 1.35 \times 10^{-3} \, mol \, L^{-1}$

$K_{sp} = [Pb^{2+}][I^-]^2 = (1.35 \times 10^{-3})(2.7 \times 10^{-3})^2 = 9.84 \times 10^{-9}$

15 Compound A

Compound A undergoes a reaction with water.

The mass spectrum of compound A has a molecular peak at 56, which gives the molecular mass. In the IR spectrum, the absence of broad peaks at 3230–3500 and 2500–3000 cm^{-1} show it is not an alcohol or carboxylic acid. The absence of a peak between 1680 and 1750 cm^{-1} indicates the absence of C=O. The peak at 1600 cm^{-1} indicates there may be a double bond. The ^1H NMR spectrum shows only two peaks, indicating only two H environments. The doublet peak at 1.5 indicates a CH and the quartet peak at 5.4 indicates at CH_3.

Compound B

B undergoes an oxidation reaction to produce C.

In the IR spectrum, the presence of a broad peak 3400–3500 cm^{-1} indicates an –OH bond and the peak around 1300 cm^{-1} could be due to C–O bond.

Compound C

Produced by the oxidation of possibly an alcohol so could be aldehyde, ketone or carboxylic acid.

The IR spectrum shows an absence of broad peaks at 3230–3500 and 2500–3000 cm^{-1}, so it is not an alcohol or carboxylic acid. The peak at 1800 cm^{-1} indicates C=O. The ^1H NMR spectrum indicates three different H environments. The triplet peak at indicates a CH_2, the quartet peak at 2.5 indicates a CH_3 and the single peak at 2.1 indicates a carbon with no H attached.

Based on the information, it is likely that:

Compound A is an alkene.

Compound B is an alcohol and compound C is a ketone.

Compound A

C_xH_{2x} $\qquad$ $12x + 2x = 56$ $\qquad$ $x = 4$

C_4H_8 butene

Butene reacts with water to produce butanol and butanol is oxidised to butanone.

As the final product, C, is a ketone, compound B must be a secondary alcohol – butan-2-ol.

Compound A could be either but-1-ene or but-2-ene as both could produce butan-2-ol. However, the ^1H NMR spectrum of compound A shows only two peaks, so A must be but-2-ene, which is symmetrical and has only two different H environments.

Compound A is but-2-ene.

Compound B is butan-2-ol.

Compound C is butanone.

Practice examination

SECTION I PAGE 211

1 B $\quad$ **2** C $\quad$ **3** B $\quad$ **4** B (Markovnikov's rule) $\quad$ **5** C

6 B $\quad$ **7** A $\quad$ **8** A $\quad$ **9** B $\qquad\qquad\qquad\qquad\qquad$ **10** D

11 D

Working:

$$c(N_2O_4) = \frac{n}{V} = \frac{1.10}{2} = 0.55 \, mol \, L^{-1}$$

	N_2O_4	$\rightleftharpoons$	$2NO_2$

I 0.55 $\qquad\qquad\qquad$ 0

C $\dfrac{-0.90}{2} = -0.45$ $\quad + \, 0.90$

E (0.55 – 0.45 = 0.10) $\quad$ 0.9

12 C

13 A

Working:

$Fe^{3+}(aq) + 3OH^-(aq) \rightarrow Fe(OH)_3(s)$

The formation of solid $Fe(OH)_3$ reduces the concentration of $Fe^{3+}(aq)$; hence, equilibrium shifts to the left, the reactant side, to produce more $Fe^{3+}(aq)$.

14 D

15 A

Working:

pOH = 14 – pH = 14 – 9.3 = 4.7

$[OH^-] = 10^{-4.7} = 2.0 \times 10^{-5}$

Ratio Ba : OH = 1 : 2; therefore, $[Ba^{2+}] = \dfrac{1}{2} \times 2.0 \times 10^{-5} = 1.0 \times 10^{-5}$

16 B

17 A

Working:

$K_{sp} = [Ba^{2+}][OH]^2$

$2.55 \times 10^{-4} = [4.8 \times 10^{-3}][x]^2$

$\sqrt{5.3 \times 10^{-2}} = x$

$x = 2.3 \times 10^{-1} \, mol \, L^{-1}$

18 D $\qquad\qquad\qquad$ **19** B $\qquad\qquad\qquad$ **20** D

SECTION II PAGE 216

21 $2HCl(aq) + Ba(OH)_2(aq) \rightleftharpoons BaCl_2(aq) + H_2O(l)$

$n(HCl) = c \times V = 0.220 \times 0.042 = 0.009\,24 \, moles$

$n(H^+) = 0.009\,24 \, moles$

$n(BaOH) = c \times V = 0.1 \times 0.055 = 0.0055 \, moles$

$n(OH^-) = 2 \times 0.0055 = 0.011 \, mol$

Excess $OH^- = 0.011 - 0.009\,24 = 0.001\,76 \, moles$

$c(OH^-) = \dfrac{n}{V} = \dfrac{0.001\,76}{0.042 + 0.055} = 0.0181 \, mol \, L^{-1}$

pOH = $-\log10[OH^-] = -\log(0.0181) = 1.74$

pH = 14 – 1.74 = 12.26

9780170449656

Criteria	Mark
Correct answer calculated with full working shown, including a balanced equation	3
Substantially correct or correct answer calculated with some working shown	2
At least one step included in the calculation of pH	1

22 a

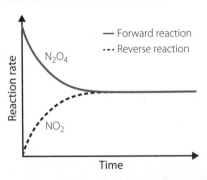

Criteria	Mark
Correct curve shapes given, axes labels used and curves labelled	2
Substantially correct graph sketched	1

b The rate of forward reaction is initially high as there is a high concentration of reactant particles that can collide, leading to successful collisions in the correct orientation to overcome E_a. The rate of the reverse reaction eventually increases as the concentration of products increases due to the initial high rate of forward reaction. Eventually, as equilibrium is reached, the rate of the forward reaction equals the rate of the reverse reaction.

Criteria	Mark
Extensive accounting for shape of curves, both forward and reverse directions, and formation of equilibrium, using collision theory	3
Collision theory used to account for some of the shapes seen in the graph	2
Some relevant statement made	1

c Brown colour becomes darker brown. Equilibrium shifts to the right, the product side, i.e. the endothermic direction to use up the heat because according to Le Chatelier's principle when a change is made to a system at equilibrium, the system adjusts in such a way as to counteract the change.

Criteria	Mark
Correct explanation of colour change using some Le Châtelier's principle theory	3
Substantially correct explanation of colour change using some Le Châtelier's principle theory	2
Some relevant information given	1

23 a $HO—CH_2—CH_2—OH$
Ethylene glycol

$HO—\overset{\displaystyle O}{\underset{\displaystyle ||}{C}}—\bigcirc—\overset{\displaystyle O}{\underset{\displaystyle ||}{C}}—OH$
Terephthalic acid

Criteria	Mark
Two correct monomers drawn	2
One substantially correct monomer drawn	1

b Condensation polymerisation

Criteria	Mark
Correct polymerisation given	1

c PET, being a thermoplastic, can be repeatedly melted and reshaped and pressed into moulds to make bottles or furniture, or it can be spun to form fibres for use in carpets, clothing or quilts. When the product reaches the end of its useful life, PET can be recycled to make clothing, carpet, luggage, bottles, tote bags, and food and beverage containers.

Criteria	Mark
At least two uses of PET outlined and specifically linked to structure and/or thermoset nature	3
At least one use of PET outlined and specifically linked to structure or thermoset nature OR Two uses of PET described with some attempt to link to structure or thermoset nature	2
Some relevant information given	1

24 $K_{eq} = \dfrac{[SO_3]^2}{[SO_2]^2[O_2]}$

	$2SO_2(g)$	$+\ \ O_2(g)$	$\rightleftharpoons\ \ 2SO_3(g)$
Initial concentration	4.70	3.21	0.00
Change in concentration	$-(2 \times 1.86)$ $= -3.72$	$-(3.21 - 1.35)$ $= -1.86$	$+(2 \times 1.86)$ $= +3.72$
Equilibrium concentration	0.98	1.35	3.72

$K_{eq} = \dfrac{[SO_3]^2}{[SO_2]^2[O_2]} = \dfrac{[3.72]^2}{[0.98]^2[1.35]} = 10.673 = 10.7 \text{ (3 sig fig)}$ w

Criteria	Mark
K value calculated showing full working, including K expression	3
K value substantially correct, or showing most working	2
At least one correct step shown towards calculating K	1

25 a The pH values of bases depend on the strength of the base and ionisation of hydroxide ions. As sodium hydroxide is a strong base, it will completely ionise to produce sodium and hydroxide ions $NaOH(s) \rightleftharpoons Na^+(aq) + OH^-(aq)$, whereas because ammonia is a weak base, only some of the ammonia molecules accept a proton from water $NH_3 + H_2O \rightleftharpoons NH_4^+ + OH^-$. As pH = 14 – pOH, a base that completely ionises, therefore, having a high hydroxide concentration would have a high pH, whereas a weak base where there is a lower hydroxide concentration would have a lower pH.

Criteria	Mark
Thorough explanation of pH difference using at least one equation and identification of substances as weak and strong bases	3
Sound explanation of pH difference using idea of base strength	2
Some relevant information	1

b The Brønsted–Lowry theory classifies a base as a proton acceptor. An Arrhenius base is a hydroxide – and forms hydroxide ions as the only negative ions in solution. Sodium hydroxide produces OH^- ions in solution, as defined by Arrhenius. These hydroxide ions act as a proton acceptor so satisfy the Brønsted–Lowry theory. Ammonia, however, does not produce hydroxide ions in solution as defined by Arrhenius but acts as a Brønsted–Lowry base in that it accepts a proton from water.

Criteria	Mark
Answer shows comprehensive understanding of both theories, with correct use of examples given	3
Answer shows some understanding of both theories, with limited ability to use examples given	2
Some relevant information about one theory or one compound	1

26 a When ammonia is added, it reacts with Ag^+ ions to form the compound $Ag(NH_3)_2^+(aq)$, thus reducing their concentration in solution. According to Le Chatelier's principle, some $AgCl(s)$ would dissociate (first reaction would go to the left) to increase the concentration of Ag^+. If enough ammonia is added, all of the solid will dissolve as first reaction will continue to go to the left to replace Ag^+ ions.

Criteria	Mark
Correct explanation including Le Chatelier's principle	2
Answer shows some understanding	1

b The addition of nitric acid (HNO_3) would cause the following reaction with ammonia:

$$HNO_3(aq) + NH_3(aq) \rightarrow NH_4^+(aq) + NO_3^-(aq)$$

This would result in a decrease in the concentration of NH_3. According to Le Chatelier's principle, the second reaction would move to the left to increase the concentration of NH_3 and this would also increase the concentration of Ag^+. The increase in the concentration of Ag^+ would then, according to Le Chatelier's principle, cause the first reaction to move to the right, producing $AgCl(s)$ and decreasing the Ag^+ concentration.

Criteria	Mark
Through explanation given using both reactions and Le Chatelier's principle	3
Sound explanation, Le Chatelier's principle mentioned	2
Some relevant information	1

27 pH at equivalence = 9; therefore, volume = 22.0 mL of NaOH required.

Equivalence point = 22 mL; therefore, half the acid neutralised at 11 mL = pH 5

At half equivalence point, pH = pK_a; therefore, $pK_a = 5$.

$K_a = 1 \times 10^{-pK_a} = 1 \times 10^{-5}$

Discrepancies between the experimental and theoretical values may occur due to:

◗ systematic error due to miscalibration of the pH probe used to record the titration curve

◗ random error in measurements of primary and secondary standard solutions

◗ student error in incorrect handling of the pipette, failure to correct rinse the burette or conical flask.

Criteria	Mark
Titration curve annotated to show equivalence point, value used correctly to calculate K_a with full working and determine accuracy	4
An equivalence point determined and used to calculate K_a with minor errors and determine accuracy	3
Some correct steps in the calculation of K_a	2
At least one correct step in the calculation of K_a	1

28 a Multiply step 2 equation by 3:
$6SO_2(g) + 3O_2(g) \rightleftharpoons \cancel{6SO_3(g)}$
Multiply step 3 equation by 2:
$\cancel{6SO_3(g)} + 2H_2SO_4(l) \rightarrow \cancel{2H_2S_2O_7(l)}$
Multiple step 4 equation by 2:
$\cancel{2H_2S_2O_7(l)} + 2H_2O(l) \rightarrow 4H_2SO_4(l)$
Add the equations, cancelling equal species on each side:
$6SO_2(g) + 3O_2(g) + 2H_2O(l) \rightarrow 2H_2SO_4(l)$

Criteria	Mark
Correct final equation written, showing full working	2
Final equation correct without any working OR Substantially correct equation written with minor error in shown working	1

b According to the equation for step 2, the stoichiometric ratio of the reactants is $2SO_2:1O_2$. A 1:1 ratio would give an excess of O_2 gas. This would increase the efficiency of the process in two ways. Firstly, an increased concentration of oxygen would increase the rate of the forward reaction because there would be an increased number of collisions. Secondly, according to Le Châtelier's principle, increasing the concentration of a reactant would increase the rate of the forward reaction, therefore increasing the yield of sulfur trioxide.

Criteria	Mark
Extensive explanation given using both equilibrium and reaction rate concepts and data from equation	4
Thorough explanation given involving both equilibrium and reaction rate concepts	3
Thorough explanation of equilibrium OR reaction rate OR Sound explanation given involving both equilibrium and reaction rate	2
Some relevant information given	1

c Using the equation for answer to part a:

$6SO_2(g) + 3O_2(g) + 2H_2O(l) \rightarrow 2H_2SO_4(l)$

$n(SO_2):n(H_2SO_4) = 6:2$

$n(SO_2) = \dfrac{m}{M} = \dfrac{150 \times 10^3}{32.07 + (2 \times 16.00)} = 2341.2 \text{ mol}$

mole ratio $\dfrac{n(H_2SO_4)}{n(SO_2)} = \dfrac{2}{6} = \dfrac{1}{3}$

$n(H_2SO_4) = \dfrac{2341.2}{3} = 780.396 \text{ mol}$

$M(H_2SO_4) = (2 \times 1.008) + 32.07 + (4 \times 16.00) = 98.086 \text{ g mol}^{-1}$

$m(H_2SO_4) = n \times M = 780.396 \times 98.086 = 76545.9 \text{ g}$

Yield (75% efficient) $= 76545.9 \times 0.75 = 57409 \text{ g} = 57.4 \text{ kg}$ (3 sig fig)

Criteria	Mark
Correct yield calculated showing full working, units and significant figures	3
Substantially correct calculation of yield	2
At least one correct step given in the yield calculation	1

29

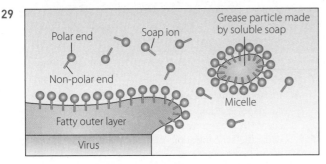

Soaps are long-chain molecules that have hydrophobic hydrocarbon tails and polar or negatively charged hydrophilic heads.

▶ The non-polar end of the soap ions is able to clean hands and remove the virus because it can bond with the non-polar fatty outer layer of the virus through dispersion forces.

▶ The ionic end bonds to water molecules through ion–dipole forces.

▶ The virus on the hands are loosened by rubbing hands during the washing process.

▶ The non-polar end of the soap ions surrounds more and more of each fatty outer layer of the virus as they are loosened from the hands by further rubbing.

▶ This results in the formation of micelles.

▶ The virus particles are dispersed throughout the water in the form of an emulsion whereby the fatty outer layer is surrounded by the non-polar end of soap ions, with the polar heads bonded to water through ion–dipole interactions.

Criteria	Mark
Soap action extensively described in terms of soap ion structure, micelle formation, process of cleaning and intermolecular forces	5
Soap action thoroughly and mostly correctly described in terms of the points above	4
Soap action substantially correct and described in terms of some features above	3
Some features of soap action described	2
Some relevant information presented	1

30 $2NaOH(aq) + CO_2(s) \rightarrow Na_2CO_3 + H_2O$

$n(NaOH) = c \times V = 1.50 \times 0.100 = 0.150 \text{ moles}$

$n(HCl) = c \times V = 1.50 \times 0.0328 = 0.0492 \text{ moles}$ used to titrate excess NaOH

Therefore, moles(NaOH) used to neutralise $CO_2 = 0.150 - 0.0492 = 0.1008 \text{ moles}$

Molar ratio $NaOH:CO_2 = 2:1$; therefore,

$n(CO_2) = \dfrac{1}{2} \times 0.1008 = 0.0504 \text{ moles}$

Mass $(CO_2) = n \times M = 0.0504 \times 44.01 = 2.218 \text{ g}$

% purity $= \dfrac{2.218}{3.00} \times 100 = 73.9\%$

Criteria	Mark
Correct calculation of percentage purity with full working shown, two balanced equations and units shown throughout	5
Substantially correct calculation with mostly full working and at least one equation	3–4
Some correct steps shown in the calculation	2
At least one correct step shown in the calculation	1

31

| Propanal | Propanone | Prop-2-en-1-ol |

The mass spectrum has a base peak at $m/z = 29$. This is the most abundant fragment and is most likely to be $[CH_3CH_2]^+$ and/or $[CHO]^+$. Neither of the other two isomers can readily produce these fragments.

The ^{13}C NMR spectrum shows a compound with three carbon environments, thus eliminating propanone, which has two carbon environments. Prop-2-en-1-ol also has three carbon environments; however, the peak just above 200 ppm is consistent with the presence of an aldehyde. There are no peaks in the ranges 90–150 ppm or 50–90 ppm, which would indicate a double bond and an alcohol respectively.

The ^1H NMR spectrum shows three hydrogen environments. This also eliminates propanone, which would show only one peak as all the H atoms are equivalent and prop-2-en-1-ol, which has five H environments. The spectrum confirms the compound is propanal by the peaks at 2.5 and 1.1 ppm, indicating the presence of CH_3 and CH_2 groups. Therefore, the correct structure is propanal.

Criteria	Mark
Gives correct structure and justifies extensively using all spectra	7
Gives correct structure and justifies thoroughly using all spectra OR Gives substantially correct structure and shows extensive justification – with an error	6
Gives substantially correct structure and justifies thoroughly using at least two spectra	4–5
Gives a structure with some correct features and provides some justification for the features presented	2–3
Provides some relevant information	1

32

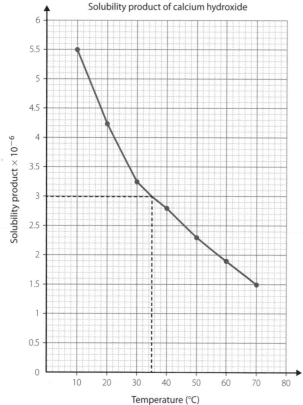

Solubility product of calcium hydroxide

$Ca(OH)_2(s) \rightleftharpoons Ca^{2+}(aq) + 2OH^-(aq)$
At 35°C, $K_{sp} = 2.96 \times 10^{-6}$
$K_{sp} = [Ca^{2+}][OH^-]^2 = s \times (2s)^2 = 4s^3$
$2.96 \times 10^{-6} = 4s^3$

$s = \sqrt[3]{\dfrac{2.96 \times 10^{-6}}{4}} = 9.0450... \times 10^{-3}\,mol\,L^{-1}$

At 35°C, $c(Ca(OH)_2) = 9.0450... \times 10^{-3}\,mol\,L^{-1}$
In 1 L, $n = 9.0450... \times 10^{-3}\,mol\,L^{-1}$

$m = nM = 9.0450... \times 10^{-3} \times (40.08 + 2 \times (16.00 + 1.008))$
$= 9.0450... \times 10^{-3} \times 74.096 = 0.670\,g\,L^{-1}$
So, solubility is $0.0670\,g/100\,mL$.

Criteria	Mark
Extensively detailed graph constructed with axes labels, units, appropriate scale, accurately plotted points, curve of best fit Graph interpolated to determine solubility product Value used to calculate solubility, showing full working and units	6
Substantially correct graph with minor errors, interpolation and value used to correctly calculate solubility OR Correct graph, interpolation and value used to calculate solubility with a minor error	5
Substantially correct graph and calculations used to calculate solubility with a key omission/error or several minor errors	4
Several correct steps, and/or substantially correct graph used to show some understanding of the calculation of solubility	3
Some correct steps shown OR some correct construction of the graph	2
Some relevant information presented	1

33 a $c(H_3C_6H_5O_7) = c(H_2C_6H_5O_7^-) = 0.350\,mol\,L^{-1}$
$V(H_2C_6H_5O_7^-) = 55.0\,mL$
$n(H_2C_6H_5O_7^-) = c \times V = 0.350 \times 0.055 = 0.019\,25\,mol$
$M(NaH_2C_6H_5O_7) = 22.99 + 7 \times 1.008 + 6 \times 12.01 + 7 \times 16.00$
$= 214.106$
$m(NaH_2C_6H_5O_7) = n \times M = 0.019\,25 \times 214.106 = 4.12\,g$

Criteria	Mark
Correct mass calculated showing full working	2
Some correct steps included towards calculation of mass	1

b Increasing the concentration of the buffer means increasing the concentrations of both citric acid and monosodium citrate by the same amount as buffers are most effective when the concentrations of the acid and its conjugate base are the same.
$H_3C_6H_5O_7(aq) + OH^-(aq) \rightleftharpoons H_2C_6H_5O_7^-(aq) + H_2O(l)$
$H_2C_6H_5O_7^-(aq) + H_3O^+(aq) \rightleftharpoons H_3C_6H_5O_7(aq) + H_2O(l)$
Increased concentrations of citrate and citrate ions will be able to react with more $[H_3O^+]$ or $[OH^-]$ ions added to the system, thereby, further increasing the capacity of the buffer solution.

9780170449656

Criteria	Mark
Le Chatelier's principle used along with balanced equation to extensively explain BOTH system adjustments when acids and bases added	4
Le Chatelier's principle used to thoroughly explain both system adjustments when acids and bases added	3
Le Chatelier's principle used to soundly explain at least one system adjustment when acids and bases are added	2
Some relevant theory included	1

34 **1** Ethanoic acid can be identified by use of sodium carbonate

2 Ethanol by use of acidified dichromate

3 Pentane and pentene by use of bromine water

Answer must have a risk assessment for bromine and dichromate, must have equations, and must include all observations made (bubbles or colour changes).

Criteria	Mark
Logical method presented to correctly identify all substances, including risk assessment, volumes, chemicals used, observations made and conclusions about identification fully explained At least three chemical equations included, with all organic structures drawn in full expanded style	7
Logical method presented with all features above included, and most substantially correct, all substances correctly identified and conclusions explained At least two equations used	5–6
Substantially logical method presented with many features above present substantially correct, at least two substances correctly identified and some conclusions explained At least one equation used	3–4
A substantially correct method used to identify at least one substance, with minimal detail included	2
Some relevant information used	1